CANADIAN MEDICAL LIVES NO. 9

EARLE P. SCARLETT
A Study in Scarlett

F.W. MUSSELWHITE
Series Editor: T.P. Morley

Hannah Institute
&
Dundurn Press
Toronto and Oxford
1991

Design and Production: JAQ
Copy Editor: Judith Turnbull

Dundurn Press wishes to acknowledge the generous assistance and ongoing support of The Canada Council, The Book Publishing Industry Development Programme of the Department of Communications and The Ontario Arts Council.

Care has been taken to trace the ownership of copyright material used in the text, including the illustrations. The author and publisher welcome any information enabling them to rectify any reference or credit in subsequent editions.

J. Kirk Howard, Publisher

Canadian Cataloguing in Publication Data

Musselwhite, F.W.
  Earle P. Scarlett

(Canadian medical lives ; no. 9)
Co-published by the Hannah Institute for the History of Medicine.
Includes bibliographical references and index.
ISBN 1-55002-096-X

1. Scarlett, Earle P., 1896–1982. 2. Physicians –
Canada – Biography. I. Hannah Institute for the
History of Medicine. II. Title. III. Series.

R464.S33M8 1991   610'.92   C91-094670-1

Dundurn Press Limited
2181 Queen Street East
Suite 301
Toronto, Canada
M4E 1E5

Dundurn Distribution Limited
73 Lime Walk
Headington
Oxford, England
OX3 7AD

CANADIAN MEDICAL LIVES SERIES

The story of the Hannah Institute for the History of Medicine has been told by John B. Neilson and G.R. Paterson in *Associated Medical Services, Incorporated: A History* (1987). With the creation of the Institute the AMS endowed it with funds, capability and responsibility to develop and disseminate a body of medical history hitherto hidden in the country's memory.

A grants and personnel support program enables the Institute through medical, social and political historians to become a powerful influence in the rapidly expanding interest accorded to medical history. A chair in the history of medicine is supported at each of the five Ontario medical schools. Originally the Institute's support was limited to Ontario, but it has now been extended to medical historical work in any part of Canada.

Earle Scarlett's contribution to medicine, and therefore to the public, was proclaimed through a stream of essays in the medical literature. With his pen, he struggled to humanize a profession he saw surrendering its virtues of scholarship, compassion and selflessness to a brand of scientific materialism. Bill Musselwhite, who combines the two professions of journalism and the priesthood, is well equipped to produce this study of Scarlett.

Future volumes in the series include Maude Abbott (Douglas Waugh), R.G. Ferguson (C. Stuart Houston), J.C.B. Grant (C.L.N. Robinson), Harold Griffith (Richard Bodman and Deirdre Gillies) and Mahlon W. Locke (Robert Jackson).

There is no shortage of meritorious subjects. Willing and capable authors are difficult to acquire. The Institute is therefore deeply grateful to the early authors who have already committed their time and skill to the series.

T.P. Morley
Series Editor
1991

CANADIAN MEDICAL LIVES SERIES

* published

CANADIAN MEDICAL LIVES NO. 9

EARLE P. SCARLETT

A Study in Scarlett

F.W. MUSSELWHITE

Series Editor: T.P. Morley

CONTENTS

To Sherry and Kim
with love

ACKNOWLEDGMENTS

ALTHOUGH ANY BLAME IS MINE, the credit for this book goes to many people. At the top of the list is, of course, Dr. J.S. Gardner, a man who combined knowledge and wisdom with a great heart and an appreciation for the wonder and beauty of life. He was one of the most loved members of the Calgary medical community, a man who in his mid-eighties kept up an active life people half his age could envy. Shortly after I gave him the finished manuscript to read, he died in a skiing accident. Given his role in it, this book must be considered a part, albeit a very small one, of his legacy to the medical community.

My thanks must also go to the Hannah Institute, Dr. Scarlett's family, the Glenbow Foundation and the University of Calgary Medical Library and its staff. W.B. Bean, M.D., who holds the titles of Sir William Osler Professor and Emeritus Professor of Medicine at the University of Iowa, has my gratitude for allowing me access to his correspondence with Dr. Scarlett.

This tribute could not have been written without the spade work done by medical historian Charles Roland, M.D., nor without the help, support, patience and better judgment of my wife, Sherry, and my daughter, Kimberly.

Material cited is, in general, available to the public. Other material quoted, but not cited, is from personal interviews, letters, unprinted manuscripts and other sources. In the Notes section, "Roland" refers to the manuscript of a lengthy interview with Dr. Scarlett by Charles Roland in 1978; this manuscript forms part of the Scarlett Collection of the University of Calgary Medical Library Historical Archives. Professor Bean has plans to make the Bean/Scarlett letters available to the public. It is a remarkable collection and I hope this will happen.

Unless otherwise indicated, all photographs were provided by Betty Dorotich, Scarlett's daughter. I extend my sincere thanks.

Note: The epigraphs at the head of each chapter are taken from various stories in the Sherlockian canon by Sir Arthur Conan Doyle – or rather, as Dr. Scarlett would no doubt say, by John Watson, M.D., assisted by Sir Arthur as his literary agent.

INTRODUCTION

"I have some papers here," said my friend Sherlock
Holmes, as we sat one winter's night on either side of the
fire, "which I really think, Watson, it would be worth your
while to glance over."
          – "The Adventure of the Gloria Scott"

Substitute J.S. (Smitty) Gardner, M.D., for Sherlock Holmes in the above quotation, and that's approximately the way this book came to be. In this case, the box holding the papers was not japanned, as it was in Conan Doyle's story, but was made of cardboard and was overflowing with letters, journal reprints and memorabilia pertaining to one Earle Parkhill Scarlett. Smitty Gardner, a well-known Calgary surgeon and patron saint of the University of Calgary Medical School, had been asked by the Hannah Institute for the History of Medicine to write a biography of Dr. Scarlett. His suggestion was that I, armed with that box of papers, take on the job.

After some thought, and feeling completely overwhelmed at the prospect of the task, I agreed to play Dr. Scarlett's Watson.

I first met Dr. Scarlett in the early 1970s when, as a reporter for the *Calgary Herald*, I decided to write a feature on that strange group of Sherlock Holmes enthusiasts known as the Baker Street Irregulars. Although something of a fan of the famous sleuth myself, I didn't know if there were any members in Calgary who could help me in my research. I had, however, spotted a listing for a Dr. Moriarty in the phone book and figured that any man carrying the name of Holmes's arch enemy must be an Irregular.

It was a good guess. Though Dr. Moriarty had ceased to be an active member, he had a suggestion. The man I really ought to talk to, he said, was a retired internist named Dr. Earle Scarlett. Scarlett, apparently, had dragooned Moriarty into the Irregulars and was one of Canada's most enthusiastic Sherlockians.

A few days later I found myself at 409 Roxboro Road. Dr. Scarlett, a medium-sized man, ushered me into a slightly messy but cheerful, book-lined study. He offered me a seat, then sank into an easy chair and lit a sturdy-looking pipe. While the smoke censed – some of his friends would say "stenched" – the room, he talked and I listened. I soon realized I was in the presence of a remarkable and erudite physician whose wide-ranging mind did not stray from medicine as much as draw other subjects into that discipline.

In later smoke-filled conversations in Dr. Scarlett's study, he talked of myths and medicine, of music and morality, of art and literature. It is difficult to think of a topic of which he was ignorant. If I mentioned a person, such as Sherlockian Vince Starrett, he would turn out to be a friend of Scarlett's, or at least a correspondent. If I mentioned a place in Europe, it usually transpired either that he had been there and knew historians or archaeologists who were studying the area or that he just happened to be a member of some group devoted to tending an associated literary grave.

It wasn't that he bragged; he simply enjoyed talking. As his daughter Elizabeth said:

> He loved to talk. He loved being at the centre of things. Father seemed to know everyone in the world. He cherished his correspondents – he would talk about those people – he always felt humbled in some ways. He had a fascination with genius – what is genius and why it occurs. He talked so much about exceptional people: "the great" . . . He cherished being in touch with people who believed in the things that matter as he would so often say.

His correspondents, from well-known literary and musical figures to people he chanced to meet on his trips, seemed to cherish him in return. When Earle and Jean Scarlett's friends on the Island of Mull died of old age, the letters, now written by their friends' children, kept on coming. The Scarletts had visited Mull partly because it was in Scotland – and the doctor's anglophilia was only surpassed by his love for things Scottish, particularly oatmeal porridge – and partly because Calgary is named after a village on the island. When the *Herald* (where I work) made a mistake about the location of the original Calgary or the meaning of the word, I could expect a phone call the next day to ensure the guilty party was corrected.

During conversations in his study, he would delve deeper into the faults of various writers from the *Herald* or any other periodical. He was especially impatient with the level of English used. Scarlett's devotion to medicine was almost equalled by his intense love for the English language and his sorrow over the depths to which he feared it was sinking. This devotion showed up in his own writing and in his voracious appetite for the printed word. His study was edged with battlements of books and guarded by stacks of magazines and clippings. There were books in the front hall, books in the living room, and books in what he called his library "stacks" in the basement. My wife, studying for her university degree, needed a copy of Fraser's *The Golden Bough*. Not wanting to trek across the city to the university library, I asked Dr. Scarlett if he had a copy. "Which would you prefer," he asked politely, "the one- or two-volume edition?" He left the room and a few minutes later returned with the requested book. Somehow I wasn't surprised.

Few people saw his library stacks. This *sanctum sanctorum* took up much of the basement of his comfortable two-storey house. While his son was at home, the basement was often littered with chemicals and electronic parts and occasionally clouded in smoke (on at least one occasion the fire department had to be called in), but in later years it was a library equipped in Spartan fashion with a chair, a light and a small desk. That he could keep track of the whereabouts of a certain volume is a tribute to his excellent memory; he was a general in an intellectual army and his stacks enclosed the main ranks. After Scarlett's death, his family kindly invited me to go downstairs and take a few books as keepsakes. Walking about among his silent companions made me nervous – I sensed I was invading the old physician's privacy and I was not at all sure he was happy at my being there.

Scarlett was known as a bibliophile but he sometimes rejected that title. "Do not imagine . . . that I place any unique or surpassing value on books as such," he wrote in an unpublished manuscript. "I trust that I am no myopic or grubby bibliophile. Books are among the happier and more rewarding amenities of life. But they should be enjoyed lustily like the other good things."

Another of the "good things" that mattered was music, especially the music of Mozart. In an appreciation of Scarlett in the first number of the *Canadian Bulletin of Medical History*, Dr. Gardner quoted Scarlett: "The music of Mozart has been one of the greatest gifts of this life." While I knew that Scarlett loved music and that he and his wife Jean played the recorder and performed baroque music in concerts around the city, I was unaware of just how deep that interest was until I learned from my mother, Florence Musselwhite, that Dr. Scarlett was known in Europe as a musicologist. He is, in fact, cited as such in *The Canadian Encyclopaedia of Music*.

His reputation stemmed from columns he wrote on such notables as Bach, Handel and Mozart, columns which, mixing medicine and music, were avidly read by musicians and composers as well as physicians and surgeons.

It was through his columns in the *Archives of Internal Medicine*, as the "Doctor Out of Zebulun," that he gained his widest audience, for Scarlett, in his later years, was best known as a writer. He did not, however, take up the pen as a way of filling in time after his retirement. His association with printer's ink goes back to his college days in Winnipeg, when he dallied with part-time journalism at the *Winnipeg Free Press*. Along the way he helped initiate two journals, including one of the few North American publications on medical history, and contributed to many more. His columns in both the *Archives* and *Group Practice* were important to him as a means of passing down, not technical knowledge, but the cultural inheritance he believed should be part of a doctor's soul.

Narrow-thinking annoyed him; in a physician it disheartened him. When Calgary's medical school took shape in the 1970s, he would quiz his friend Smitty Gardner, a member of the admissions committee, about the would-be doctors. He would ask in a worried voice, "They're not just in it for the money are they, Smitty?" Dr. Gardner would reply that if he thought any student was "just in it for the money" that student would be refused admittance. It was an assurance the ageing physician needed.

This small book is a tribute to a man who was a doctor first but who saw medicine not merely as a science but as one of the greatest of the humanities. Scarlett believed doctors should realize they were the heirs of two worlds: the world of Hippocrates, Galen and Harvey, and that of the arts, history and literature. As the title suggests, this is a "study" rather than a formal biography. I hope the inclusion of many of Scarlett's own words will move the reader to seek out more of his writing. It is easily accessible in a variety of journals and in Dr. Charles Roland's anthology, *In Sickness and in Health*.

I have attempted to show Dr. Scarlett as he was, his failings as well as his virtues. I may be accused of trying to paint too flattering a picture. I hope I have not done so. He had his critics, and he was not admired by all his colleagues in this city. Calgary's medical community, like that of most cities, has always been divided along several lines, and Scarlett was very much a member of one particular group.

To some he seemed aloof but as Dr. Gardner insisted, this was not an attempt to distance himself from others; it was the mark of a "shy reserve and a preoccupation with his own busy world."

As he grew older he was at odds with his time, a Victorian who never lost the sensibilities of an age which had passed even when he was a child. I am proud to have known him.

Dr. Earle P. Scarlett.

Chapter One

He had the weather-beaten appearance of one who
has spent most of his time in the open air, and yet there
was something in his steady eye and quiet assurance
which indicated the gentleman.
— *Hound of the Baskervilles*

EARLE PARKHILL SCARLETT WAS A child of the Canadian prairie. Not the modern prairie with its high-speed roads, but the pioneer world of oxen, wheat rust and the prairie's one saviour from isolation, the railroad. If this helped shape his character, so did one other factor. Like Sir William Osler, the man whom Scarlett called "his medical idol," he was a child of the manse.

Throughout his life he remained a man of his era, a Victorian, born into the Protestant tradition which called for a man to stand on his own feet and seek his own fortune. He was born at a time when English-speaking Canadians looked back to Great Britain as the guardian of culture, a term which covered everything from oatmeal porridge to great literature. He shared this attitude, although he eschewed the all too common British Protestant dislike and distrust of French-speaking Roman Catholics; this part of his heritage he discarded while fighting beside French Canada's Royal 22nd Regiment during the First World War.

The parsonage that was his first home was a tiny shack in High Bluff, Manitoba, a fly speck on the map of the Canadian Pacific Railroad west of Winnipeg. It was, in fact, so small that when his mother came close to term she was moved to a farmhouse owned by a man named Dillworth, a house thought to be more suitable than the tiny manse for a woman in labour. There, on 27 June 1896, she gave birth to twins. Only one, a boy, survived. He was given the name of Earle Parkhill.

The father, Rev. Robert Arthur Scarlett, may have been overjoyed, but he must have wondered what life held for a boy raised on the bald prairie far from the civilized East where he himself had been born. A Methodist minister, Robert Scarlett was a druggist from Cobourg, Ontario, who had switched from dispensing drugs to dispensing the Gospel, and this area, near the tip of Lake Winnipeg, offered little in the shape of culture.

The son of the inspector of schools for the counties of Northumberland and Durham, who came to Canada in the 1840s during the Irish potato famine, Robert Scarlett took his secondary schooling in Cobourg. He then attended Victoria University, an institution founded by one of Canada's most noted educators, Egerton Ryerson. Deciding to be a pharmacist, he went to Toronto where he obtained his degree from the School of Pharmacy, at that time on Gerrard Street.

With his parchment in hand, the young man moved to Oshawa and opened what Earle Scarlett believed was the first drugstore in the city. Robert found a place to live above the thriving blacksmith shop of a man named Sam McLaughlin, who would eventually turn his talents from horses to horseless carriages.

Robert Scarlett, away from family and friends, was a devout young man whose life outside business hours tended to revolve around his church – so much so that his commitment to pharmacy waned as his interest in religion grew. Deciding that he had a call to the Methodist ministry, he sold his shop and studied theology at Victoria College, a bastion of Upper Canada Methodism.

Earle Scarlett was born only eleven years after Louis Riel had been hanged, and a good deal of the Canadian prairie was still an empty space stretching from the Rockies to the Canadian Shield. It was, however, becoming a mecca for settlers, especially from Great Britain. The huge influx of Europeans was still in the future. The Methodist church, like other Protestant and Catholic denominations, had established missions for both settlers and native peoples, and Robert was sent to Manitoba as a student minister. After one year there, and a trip east for more training, he was posted to High Bluff, with responsibilities for Poplar Point, five miles to the east, and a nearby Sioux Indian reserve.

Earle Scarlett's mother, Alma Edith Parkhill, came from an Ontario family, notable partly because her father was a power in the Conservative Party and a friend of John A. Macdonald, but even more so because of his position in the Orange Lodge. Named after William of Orange, the Dutch prince who as King of England defeated the Irish Catholics in 1690, the Orange Lodge had imported Protestant British animosity towards the papacy into Upper Canada. During the nineteenth and early twentieth centuries, it wielded a political power modern Canadians would find hard to believe.

This stalwart Ulsterman, born on the border of the counties of Fermanagh and Cavan, told his grandson that his first recollections were of being hidden in a hedge by his mother to prevent his being murdered by Roman Catholics. Whether the story was true or not, he became a prime mover in the Orange Lodge, sharing its anti–Roman Catholic (and anti–French Canadian) sentiments.

Earle Scarlett said later that even in his student days in the 1920s "you couldn't get anywhere in Toronto unless you had certain affiliations – you couldn't get elected to Parliament or Toronto City Council except on these terms – you had to be an Orangeman, you had to be a Methodist, and you had to be a Conservative."[1]

Earle therefore grew up in a family in which a sense of the religious struggle in Ireland was a living thing. It was one legacy, however, that he did not accept. "Father was proud yet saddened by being Irish," his eldest daughter, Elizabeth, says. Her sister Katherine agrees: "The Irishness of the family filtered through in many ways but I don't think Father was much interested in the political concerns of his father."

Scarlett put it another way, saying his Irish blood on both sides of his family explains "a romantic streak which cannot be suppressed and the inclination on all occasions to put an herbaceous border around reality."

Alma Parkhill was raised in Midland, Ontario, and since her father's political and lodge connections required him to travel a great deal, she was sent to a young ladies' finishing school, Alma College in St. Thomas, Ontario. There, she learned the necessities of life for a Victorian lady: embroidery, music, singing, elocution and needlework. "All of which," her son wryly observed, "was a magnificent preparation for life in the tough settler's West."[2]

While Robert was back in the East, he met and, in August of 1895, married Alma in Midland. The bridegroom's gift to the bride was practical and caused some comment in the local newspaper. The report of the wedding noted that the bride received "a long coat, two huge mitts coming right up to the elbow, a cap, a special full robe, all made out of buffalo skins."

It was an unusual sight at an Ontario summer wedding, but Robert was a practical man and knew what his bride would face on the prairies.

By the time of his death in 1923, Reverend Scarlett was an esteemed and well-established religious figure in a Winnipeg that had outgrown its rustic days of Red River carts and fur traders. The 1890s, however, were still pioneer days on the prairie, and to the settlers this missionary was a man with a wife, a love of books, a young son and little else. He conducted his ministry amid some of the finest farming country in Canada, the Portage plains west of Winnipeg; Robert Scarlett, who loved horses as well as the Bible, seems to have enjoyed the trust and respect of his neighbours.

The future physician and writer would take pride in his birth in the Dillworths' prairie farmhouse. Six months before his mother bore her twins, a Dillworth son named Ira was born. Ira became a professor of English at the University of British Columbia and head of the Canadian Broadcasting Corporation. Earle and Ira became lifelong friends and formed a whimsical two-man organization, the High Bluffs Old Boys' Association, on the grounds that they were the only two children they knew of born in that area around that time. Years later, when Earle made a speech in Winnipeg, and he was introduced by Premier Douglas Campbell, who also turned out to be a child of High Bluff, Scarlett immediately offered Campbell the third membership in the Old Boys' Association, and the premier became the society's treasurer.

It was also a source of pride to Earle that he was a child of the manse. "A man can't have a better dower with which to start in the world," he said.[3] Elizabeth, however, remembered that he rarely discussed his family life at High Bluff. "I wish he had done [so] more and had written it down – for now we have just scraps of information."

When he did talk to his children about his upbringing as a minister's son, he would speak of the sense of duty his father and mother drummed into their children. If visitors came, they "would be given the biggest piece of pie while the small children watched in concealed dismay."

"His mother had to put up with the frequent moves of her husband's calling," Katherine noted, "and they always had an extra place at dinner for the numerous visitors."

"It was a frugal life," her father told her. His parents had little money for such frivolities as toys, but his father did have books. As a toddler Earle found them both fascinating and frustrating. Dr. Smitty Gardner remembers his friend saying, "I was crawling around the floor, and in a minister's house there are not very many toys. So I would pull books out of the bottom shelf and look at pictures. Books were my first playthings. There is printer's ink in my veins, I know it. I couldn't learn to read fast enough because I knew all those books were there to be read."

One of those books was the Bible. While he wasn't conventionally religious, he regarded the Bible as the greatest book ever written, and used it constantly to help illustrate a point he was making. Another of his favourites was *Pilgrim's Progress*, which he read four or five times from cover to cover, Dr. Gardner believes. Scarlett enjoyed quoting from this great novel. In 1978, on the tercentenary of its publication, Scarlett advised students attending Calgary's E.P. Scarlett High School to read it and reread it, with a pencil handy for underlining. "This book, written in prison . . . by the son of a tinker, is one of the glories of our race," he wrote. "Side by side with the Bible it helped form our language."[4]

Pilgrim's Progress was actually the second book he remembered reading. The first was *Foxe's Book of Martyrs*, a huge illustrated work and the first book he managed to drag from his father's shelves.

While his father's Methodism rubbed off on Earle at first, in later years the children attended church with their mother alone. Scarlett had faith – his writings leave no doubt of that – but it was not doctrinaire. His Christianity was of the practical sort, which is probably why he delighted in one story about his father, told to him by the pioneer Saskatchewan physician, Dr. Dan Thompson.

Thompson took medicine at the University of Edinburgh and then the University of Toronto. Equipped with his degree, he moved to the now vanished hamlet of Chater, Manitoba. His home and clinic was a one-room shack with a Hudson's Bay blanket dividing bedroom from surgery. When Reverend Scarlett arrived as a student minister, Thompson was happy to take him in. He had met Scarlett when the minister was studying pharmacy; his medical knowledge might come in handy, and two in one bed would help dispel the bitter prairie cold. Thompson was often called out on winter nights by some farmer whose wife was in labour. According to Scarlett, Thompson recalled this incident:

I'd . . . come back maybe an hour and a half, two hours, three hours, around five to six in the morning, colder than billy-be-damned, and shivering. I'd get my clothes off ready to dive into bed, and do you know what your father would do? He would move onto the cold side of the bed and let me get into bed on the warm side. "Now," he said, "why do I tell you that, Scarlett? I've lived in this world a long time. I've seen its ups and downs. I've thought a lot about it. I've thought a lot about so-called religion . . . I've pondered for years, what is a Christian? Now . . . I know and I've told you the story to illustrate it. A Christian is a man who moves over onto the cold side of the bed to let his chum have the warm side. That is my definition of a Christian."[5]

It was a definition with which Earle Scarlett was in accord.

Given the number of his father's books, Earle wondered whether Reverend Scarlett had some private source of income. "I can never remember a time when books and a teeming library were not as much a part of my world as the members of my family," he once wrote, although he added that their reading was tinged with the "theological melancholy" of a preacher's library.

Dr. Scarlett's appetite for books could be satisfied in a profession which enabled him to build a library of his own. He was also a tireless supporter of public libraries. Georgina Thompson, a research librarian at the Calgary public library, once remarked that "Dr. Scarlett eats books."

Soon after I met him, his heart problems made him housebound, except for walks about the neighbourhood with his dogs. This did not mean he was starved for new books, for both he and his wife had excellent contacts with the public library. A group of young librarians would, he said, snatch new books, set them apart and deliver them to his home every few weeks. He enjoyed this unusual privilege, one, as he said, with a very practical side. "In that way I keep up with a fair sample of contemporary books, but I don't have to be out of pocket and then find one's money is more or less wasted when one throws the book in the wastepaper basket."[6]

Each summer, until World War I, young Earle would be sent back to Ontario to stay with his paternal grandparents at their lakeside cottage on Georgian Bay. He never forgot his grandfather, a patriarchal figure with a long white beard, who would tell him about the days when he was young and had stowed away on a ship bound for Canada. His grandfather told him stories of the old country, of ". . . old, unhappy, far-off things / and battles long ago."

While these stories did not leave him personally involved with Ireland's perennial troubles, they gave him a feeling for history, something which was to become an important part of his life.

The last thing his grandfather would do in the evening, before leading his grandson up to say goodnight to his grandmother, was carefully to wind his gold pocket watch.

> The winding of the watch which was later mine and is now in the hands of my son (five generations) gave me my first and lasting sense of time. I then began to learn that you could not control time, but you could control what you did in it. And I learned something else. As I watched Grandpa wind the watch, I became in a flash one of the procession of generations.[7]

Methodist ministers did not live a settled life, as the church preferred to see them move every four years. The frequent moving left Earle with few memories of his early years. "By the time I'd get adjusted to one town, it was time to pack up and move to another one."[8] While living in Medicine Hat, he paid his first visit to Calgary to see his mother's oldest brother, a Canadian Pacific Railway superintendent. (Scarlett believed, with a certain amount of pride, that the Alberta hamlet of Parkhill was named after this official. This is not impossible, but cannot be confirmed.) Finally, in 1906, while Earle was in his last year of primary school, his father was given charge of one of Winnipeg's largest Methodist churches, and the hegira through the prairies ended.

A studious lad who preferred reading to sports, Earle must have been an unusual figure. At least he was reasonably proficient at the one sport played by everybody on the prairies: curling. And what the school curriculum of a rural school lacked, his father provided, ensuring he had a solid grounding in the classics, including Latin and Greek.

He may have wanted to be better at schoolyard sports but he was plagued by one particular problem. As he wrote years later, "All my life I have borne in silence, like most of my tribe, a singular burden – left handedness. No one ever seems to write or talk of this handicap. Victims are left to wrestle with it, although in recent years teachers no longer threaten poor children who quite naturally attempt to write with their left hand, as was the case in my youth."[9]

The burden frustrated him. Everything from parking meters to watches, he wrote, were made for the benefit of the "dexterous." Even men's underwear was adjusted for the right-hander.

> The effect of left-handedness on motor efficiency and co-ordination doubtless varies among individuals. Some find no difficulty; to others it is a tremendous handicap, particularly in playing various forms of sport. As one of the latter group, I can testify that left-handedness made me a "mug" at games. In the beginning I had a hard time to decide whether to bat left-

handed or right-handed, in curling which hand to deliver the stone with. The net result was that one never acquired any real skill with either stance.[10]

As a lover of words he enjoyed playing with the term "sinister," not to mention "ambidextrous," which, he said, carried an implied slur, as it meant being right-handed with both hands. Scarlett was not the only left-hander in his family, and because left-handedness has been associated with twins – and he was one himself – it was a subject that interested him.

His maternal grandmother sent the Scarlett children picture postcards, then becoming popular, and being left-handed she would write mirror-fashion, from right to left. Scarlett, who always found it natural to read a book starting with the back page, loved holding such cards to a mirror to enjoy the grandmotherly greeting. He also took special pleasure in watching a favourite anatomy professor draw body structures for his students and noting that, while this gentleman was artistically ambidextrous, his finest work was done with his left hand.

At the age of twelve, Earle published his first piece of writing. He made only a pittance for it but received something he valued more – a book on northern explorers. He kept the book in his library until his death.

Scarlett finished his schooling at Winnipeg's only high school, Winnipeg Collegiate Institute, graduating at the age of fifteen. At that point his father made it clear that it was time Earle began earning his own way. "My father told me that, from then on, I could have a roof over my head, but otherwise, my income, education, future were up to me."[11]

His first need was to make some money, and he took whatever job he could find. He turned his hand to clerking in stores, teaching piano, delivering messages on a bike, and working in a construction camp. Being well schooled in literature, history and the classics, he took his talents to a roadless part of northern Manitoba to teach summer school. Each day he drove a pair of oxen the seven miles from his lodgings in Lundar to the school, poking the slow beasts in the rump with a bamboo fishing pole.

His salary was munificent for those days, $40 a month, but he wanted to improve his education. When the summer ended, he enrolled at Wesley College, which later became part of the University of Manitoba. Literature and history still held his interest. Medicine wasn't even on the horizon. He naturally gravitated to the editorial office of his college's journal, the *Vox Wesleyana*.

Chapter Two

As for you, Watson, you are joining up . . .
— "His Last Bow"

Earle Scarlett was teaching summer school in the bush north of Winnipeg when Britain declared war on Germany on 4 August 1914. As Sir Wilfrid Laurier said, "When Britain is at war, Canada is at war," and few Canadians objected. Both old and young rushed to enlist. Their enthusiasm, in fact, led one group of Westerners to hijack a CPR train and force the engineer to take them to Ottawa so they could join one specific regiment.

Scarlett was as enthusiastic as any and returned to Winnipeg as quickly as possible to enlist. Before proceeding to the Winnipeg recruiting station, he stopped at his family home and told his father of his plans. Reverend Scarlett, who was also planning to join up, felt his son's plans were nonsensical and ordered him to go back up north and return to work. He could join the army but he was to finish his university career first.

Scarlett obeyed his father in that he did not sign up with the regular forces. He entered university but in 1915, at the age of nineteen, he joined the Western University Battalion, a semi-militia unit, which allowed him to carry on his classes while learning how to "port" arms and bash squares. The officers were, in fact, professors at the University of Manitoba.

After formal classes ended, military life picked up. Even for a university-based outfit, there seemed to be no time to write exams. Scarlett and three others appeared before the president of the university and asked to be allowed to receive their degrees without writing the exams. Even though the president was a captain in the same battalion, he refused to give his permission. Not willing to admit defeat – Scarlett, we know had a stubborn streak in him – the four men took their request to the General Faculty Council. That body decided to go along with their request, and they became the first students to graduate from the university without writing their final exams.

Before convocation rolled around, at which Scarlett would receive his bachelor's degree, he and many of his fellow khaki academics decided the university battalion was unlikely to see battle in the near future. Much to their officers' displeasure, they transferred to the 4th Division Cyclist Corps, which, so rumour had it, was about to go overseas. This unit was formed to act as battlefield messengers; some actually did this, although a bicycle was not the most efficient vehicle in a war fought in mud. Scarlett's new outfit was commanded not by academics but by regular soldiers, mainly Boer War veterans. Perhaps not unexpectedly, the colonel commanding the unit met his request to attend convocation with a blank stare.

"The colonel said, 'What the hell is convocation?' We tried to explain. 'Never heard of it,' the officer replied.

"I said to Herb [Jackson, another graduate], 'God, it looks to me that we're not going to get to convocation.'"

The colonel and his company sergeant-major had them step out of the room for a moment. When they returned, the colonel looked at them. "Well, we've thought it over, you fellows. We'll have the Orderly Room issue you a half-a-day pass . . . Don't think for a moment we're giving it to you for anything you've done. But that's the first goddamn type of that request we've ever had, and that's why we're giving it to you, for your originality."[1]

A photograph in the 1916 *Vox Wesleyana*, shows a handsome, tousle-haired young man looking very serious and slightly nervous in the uniform of a private. Earle Scarlett was the editor-in-chief of the yearbook, but he wouldn't see the picture of himself as a member of the college honour role or read the words of praise his successor, T.H. Nutall, wrote about him until months later, after his unit had been shipped to a camp in the south of England.

Like thousands of other Canadian boys, he suffered through the most inhuman conditions possible, but never, either in his diaries or his letters home, did he show a sign of regretting his decision to enlist. Active service brought him pain – he was seriously wounded – but it also brought him a wife and led to his decision to become a physician.

In the sophomoric humour common to college yearbooks, the *Wesleyana* noted that Earle's mother wanted him to be a preacher, his father a druggist, but he had become a "lady-killer." What he should have become, it stated, was "Mary Pickford's manager." As his diary records, he thoroughly enjoyed feminine company, although in the innocent manner befitting a clergyman's son, and he did spend a lot of his time at the movies. If he had any interest in medicine before he went to war, it's not mentioned in either his diaries or the college yearbook.

Private Scarlett left wartime Winnipeg on the last day of December 1916, and like many newly hatched Canadian soldiers, he ended up in Toronto's Exhibition Park. He bunked in the poultry building, which must have inspired jokes about chickens and white feathers. He seems to have regarded army life more as a nuisance than anything else. He saw a movie almost every day, caught a performance of *La Bohème*, took in a concert at Massey Hall and visited his relatives. The observation in his diary for 13 January was typical: "Went on guard duty at 4 p.m. Day wasted."[2]

The sixteenth of January, 1917, was a momentous day for Scarlett, although he did not realize it at the time. He visited the Toronto home of his Aunt Ella and there met a vivacious young university student named Jean Odell. Jean's diary for that day reads, "I went to Miss Scarlett's in the evening and met her nephew, Earle, who is at Exhibition Camp at the moment. Had a fine time."[3]

A fine time indeed must have been had by each, for they went out together the next day to a movie and a party with friends, and Earle did not get his date home until midnight. Naturally she received a scolding for behaviour not suitable for a young lady. Nonetheless, Miss Odell turned up to see the young soldier off when he shipped out the next day. The young woman must have wondered whether he was going to fight or read, for his kit-bag was loaded with books. Writing her from his camp on Salisbury Plain a month later, he described how the troop train had been stalled by heavy snow. "However," he wrote, "with the aid of the books you marvelled at . . . we fared very well."[4]

He sailed from Halifax on 26 January in a convoy guarded by two destroyers and a French cruiser, and promptly got seasick despite reasonably good weather. He had just about recovered when a gale struck and his painfully acquired sea legs collapsed. As if that wasn't bad enough, during the storm his ship lost both its escorts and the rest of the convoy. With thoughts of the dreaded U-boats haunting them, they spent ten days searching before they found the destroyers. Scarlett never did discover what happened to the cruiser.

He was happy to sight Ireland and then Wales on 5 February, but to add to the rough crossing, the ship ran aground in the River Mersey on its way to a blacked-out Liverpool. One day later he finally got ashore and through the military red tape. He could then calm himself with a Mary Pickford film.

Like many prairie boys, he was used to cold winters but not the damp, and the camps Canadians were sent to were cold, draughty and damp. On 11 February, although a good Methodist, he attended the Sunday service at the Anglican hut – on the pragmatic grounds that it was well heated.

During the day he and his mates discovered the questionable joy of route marches in mud. But at the same time, he didn't ignore the important things. He spent the mid-February nights reading Browning, and by the end of the month, he was deep into Thackeray's *Vanity Fair*.

The Canadian soldiers who were shipped to England had hoped they would march straight into battle, but like their compatriots in World War II, they discovered that the army didn't work that way. While the British were glad to have "colonial" reinforcements, they were suspicious of the Canadians' training. The Canadians were sent first to Salisbury Plain and then to camps in Sussex for further training. Neither area had much to offer the Canadian soldiers, and while Scarlett's diary doesn't mention this, the tedium they endured resulted in their getting a reputation for drunk and disorderly behaviour.

War, for the foot soldier, has been described as ages of boredom relieved by moments of stark terror. For the Canadians in Scarlett's draft, it's probably true that they would have welcomed any relief from the boredom, terrifying or not. "*Ex nihilo nihil*" (Out of nothing, nothing), he recorded in his diary for 2 March. "*Semper eadem. Arma virumque?*" (Always the same. Arms and the man?). The final phrase is a sarcastic comment on the opening words of Virgil's *Aeneid*: "*Arma virumque cano*" (I sing of arms and the man). Scarlett had little to sing about.

Things picked up later that month when he got leave to visit his father, who by this time was a captain in the Chaplain Service stationed in London. The two divided their time between religious and secular pursuits:

going to the theatre to see the phenomenal success Chu Chin Chou, taking in the sights, and attending a "stupendous Wesleyan mission" at the Albert Hall.

Back at the camp in Sussex, training finally began in earnest. While the first contingent of Canadians to go overseas prepared to attack Vimy Ridge, Scarlett began a "bombing" course in the use of grenades and fought mock battles in pretend trenches. The soldiers soon had another form of training thrown at them: how to handle the newest and most vicious weapon introduced in the war, poison gas.

The Germans introduced this hideous weapon on 22 April 1915 during the battle for Gravenstafel Ridge. They cracked open cylinders of chlorine, and the puzzled Canadians watched green clouds of gas drift across no-man's-land towards the French lines. The effect was devastating. The Allied high command had been warned that the Germans planned to use poison gas, and had even been told where it was to be used, but they did not believe it would happen. The effect of the gas on the lungs, skin and eyes of the unprepared French troops was immediate. Those who survived suffocation fled, leaving piles of the dead and near dead behind them. The gas also affected the Canadians who, as the German advance followed the gas, stayed and fought, staving off what could have been a serious German breakthrough.

At first, the only way the troops could protect themselves from the gas was by breathing through untreated cloth, which did little to protect the mucous membranes of their eyes, throat and lungs. By the time Scarlett began his training, the troops were equipped with rubber gas masks. These protected the eyes and respiratory tract but left the skin unguarded.

Scarlett threw his Mills bombs, practised with his gas mask, and wondered when his turn in the line would come. He wrote to Jean Odell with increasing frequency, and she replied faithfully, telling him about the active social life in Toronto and her studies at Victoria College within the University of Toronto. His father, stationed in Boulogne, occasionally returned to England and the two found some chances to talk.

His unit moved to the New Forest for more training in grenades, but despairing of seeing some sort of action, he and his friends applied to go to France as ordinary infantrymen. Most of them were successful, but Scarlett was left behind to share his hut with one other disappointed soldier. To escape the tedium, he borrowed a bicycle and one day rode to Oxford, a city which represented the culture he loved.

On 27 June he turned twenty-one, and seven days later he was chosen for an infantry draft. Instead of to France, however, the train took him to a camp near Brighton for more training, this time as a rifleman, which entailed spending hours at the butts. Parcels from Jean, often containing gingersnap cookies, which travelled well and would become a tradition in his home, buoyed his spirits. He survived the course and looked forward to action.

Canada, however, had enough soldiers with rifles. What it wanted were machine-gunners, and Scarlett was chosen. This meant more training, this time near Seaford, a town twenty miles east of Brighton on the Sussex coast.

By the end of September he had qualified on the Vickers machine gun, and having suffered lectures from one medical officer after another on the dangers of venereal disease, he at last felt ready for France and the enemy. So the army – obligingly – sent him on a revolver course, and it wasn't until the new year that he marched into Torquay and embarked for France.

On 17 January 1918, in "unspeakably lousy weather," his outfit carefully made its way across planks set in the mud to a reserve position at Souchez, just below the northern tip of Vimy Ridge, known as "The Pimple." Canadians had captured Vimy Ridge on 13 April 1917, but the Germans were still dug in only a few miles away. Scarlett's outfit spent the next few months in the front lines and survived heavy shelling and strafing from German aircraft before they moved back again into rest positions.

By good luck he was assigned to a sector with nearly dry dugouts equipped with real bunks. A package of cookies from Jean awaited him at mail call.

Otherwise, it was "mud, ad infinitum."

Jean had suggested they write in code so that he could give her a better idea of what was going on, but he didn't like the idea. After the war ended, he explained that military censors knew all about such codes and that penalties for trying it on were harsh. One of the men in his unit had been caught at it. He did what he could, however, writing on everything from bits of scrap paper to scavenged German stationary, hoping the latter "would not offend" her patriotism.

In a trench, by the light of a candle stuck in a box, "so as to show no reflection," he wrote with humour about what must have been a horror. If Jean remembered her Greek mythology, she would, he wrote, understand his reference to living in Erebus; his current residence was "a tiny hole about 30 feet underground whose measurements are about 7'x4'x4', thus making it necessary to live on all fours." He made light of it in his letters, but the reality was different. The experience left him claustrophobic and unable to bear upper berths and enclosed sleeping places. On 1 March he was pinned down in a shell hole by a sniper, but escaped unscathed, and on 4 March he rather gleefully records his "first show."

As the Germans attacked the Canadian positions at 5:40 a.m., Scarlett swung his machine gun into position. "The gun jammed after a few rounds" with a bullet stuck in the breech. "The raid was repulsed," he wrote, "after Heinie got back past the company HQ." Scarlett, once again, was lucky. He was isolated in a forward position, and near misses during the heavy shelling that accompanied the attack sent shrapnel ripping into the sandbags defending his gun emplacement. Some of his comrades were not so lucky. The German guns made direct hits on the positions to his left and right.

True to form, he did not allow the blood- and vermin-infested mud keep him from more cultural pursuits. On 12 March he recorded that "lately" he had read Coleridge's *Biographia Literaria*, Courtney's *Rosemary's Letter Book*, Ruskin's *Time and Tide* and Carlyle's *Past and Present*.

On 26 April he discovered at first hand what gas could do. During an attack his gas mask saved his eyes and lungs but produced sores on his face. "At agony with them," he managed to make his way back to a dressing station for treatment. He suffered with these sores until he was sent out for further treatment weeks later.

On 31 May, with his company resting behind the lines, Scarlett's father managed to visit him. "He is looking first rate and I was glad for the opportunity to talk over many things," the younger Scarlett recorded. Although he doesn't say so, some of those "things" must have involved friends he helped bury a few days before. Gradually, many of those who had arrived in France with him only a few months before were either being carried out by stretcher or shipped back a few miles to the cemetery.

After the day with his father, it was back to the front in time for yet another gas attack. This backwards-forwards movement was part of the British army philosophy that developed out of trench warfare. Since conditions at the front were hellish, units were brought out for a rest and clean up, which provided some protection from disease, the inevitable result of living in a slimy mixture of mud and decayed human remains.

On 3 June his company came under heavy artillery fire, and once again

Scarlett narrowly escaped. A few days later they were moved to positions in front of Neuville Vitesse in time to repulse a German attack. "Fired off eight belts of ammo," he records with some satisfaction. "Gun going great." He celebrated his twenty-second birthday in the trenches, wondering whether he would see a twenty-third.

In early July, again in a rest area, his father wangled him a two-day pass and took him to a hospital to visit wounded friends. Chaplains were important figures in military hospitals and Scarlett was given the freedom of the place. He was allowed to visit both the wards and the operating rooms. The contrast between the killing grounds on the one hand and the hospital on the other was something he talked about for the rest of his life.

His unit stayed in reserve during most of July and missed the worst of the Second Battle of the Marne, but on 30 July they were loaded onto cattle cars and taken close to Amiens. Sick, dead tired, and burdoned with heavy packs, they marched eleven kilometres to the front. Almost half of his company collapsed on the road, something Scarlett dutifully recorded in his diary. His pack must have been especially heavy, for he noted three days later that he had just read Hardy's *Under the Greenwood Tree*, Price Collier's *England and the English*, Leacock's *Frenzied Fiction*, Arnold's *Literature and Dogma*, a book on the English theatre by Beerbohm Tree and various issues of *Saturday Night, Maclean's* and the *North American Review of History*.

What Scarlett did not realize was that he was about to take part in two battles that would break the Axis spirit and lead, a few months later, to the surrender of Germany and Austria.

On 8 August the Battle of Amiens began and Scarlett was in the thick of it. Although held up from time to time by heavy machine-gun fire, the Canadians gained ground quickly. By the next day, Scarlett's unit had moved forward twenty kilometres to the hamlet of Rosières but was running into stiffening opposition. Attacks by German aircraft were answered by Scarlett, who turned his gun to the skies to defend Canadian artillery positions, but as far as he could make out in the noise and confusion, he failed to bring down any planes. What depressed him the most in the days that followed was the stench of the battlefield. Canadian cavalry had attacked the Germans in front of the position his unit had taken when it had bunkered down. The decomposing corpses of men and horses were still lying on the field.

By the nineteenth he was back at a rest area for a short period and was visited by his father. He discovered that his application to join the Royal Flying Corps had been turned down. His unit returned to the front in time for the Battle of Arras, which began in heavy rain on 26 August. Two days earlier he had written a cheerful "all's well" letter to Jean. "Once more they failed to hand me a 'blighty,'" he joked. (A "blighty" was a wound severe enough to take a man out of the line and ensure his transfer to Blighty, England.) Perhaps he should not have tempted fate. The next letter Jean received from Earle was mailed from a hospital ward and carried the stamp "Posted by a wounded soldier."

The battle began reasonably well, despite the weather, but casualties were heavy. On the second day, Scarlett's machine-gun unit was to support the Royal 22nd Regiment, the "Van Doos", in its assault on Rouvroi Ridge, a hill covered with barbed wire and protected by heavy machine-gun emplacements. The Van Doos had lost all their officers the previous day, and a Major Georges Vanier, described by Scarlett as "a cherubic youngster of 22," took command. "Intelligence sent up orders that there were no Germans ahead of us. We were to get up out of our little trenches we dug down at the bottom of the valley, and walk up the hill and take the hill . . . We never got to the top of the hill."[5]

Vanier, who later became governor general of Canada, led 640 men in the attack. An hour later all but 130 were dead or wounded. Scarlett's own machine-gun company attacked with 121 men at noon. Only 13 had survived forty-five minutes later.

A shell landed near Vanier, who fell within shouting distance of Scarlett with a wound that would cost him his leg. A few moments later another shell gave Scarlett his "blighty" and he was evacuated to England. He wrote to Jean Odell on 1 September 1918, from the 2nd Western General Hospital in Manchester:

We moved in to the line on the night of the 25th in a downpour of rain. Got in the front line about 2 a.m. and zero hour was 3 a.m. . . . That was a busy first day. The first few hours were a kind of nightmare of pitch black darkness, wire, trenches, gas and machine gun bullets.

The first night they found semi-protection from the rain, but the next night, after they had passed the Hindenberg Line, "was a dandy – we lay in a valley all night with Heinie cheerfully sniping at us from the ridge in front." The next day "the dirty work started."

Until then his platoon had been spared, but on 28 August, serious casualties to the troops "came with a rush."

Our machine gun section got mixed up in the first wave of infantry & by
the time we got to the top of the hill there were only a few of us left. Heinie
machine guns were as thick as flies and the beggars turned their field guns
down, firing point blank at us with his eighteen pounders. That gave me a
lovely sensation. But just as we got to the top of the hill I got mine. Shell
must have landed among us for when I came around some one was bind-
ing me up and telling me if I could make it to streak it out which I did
"toot sweet" with the help of a Heinie and two others.

Although severely wounded by shell splinters, the worst being a pen-
etrating wound in his neck, he considered himself lucky. The same shell
killed two and severely wounded four others – "all my chums," he wrote.
One splinter opened up the abdominal wall of his friend Tommy Drysdale.
"I tried to help Tommy hold his intestines in as best I could but by that
time I was too far gone."[6]

According to an account he wrote in 1960, two large German prisoners
carried him to an advanced dressing station. In shock from loss of blood,
he remembered very little of the next twelve hours. When he woke up, he
was in a tent hospital, known as "the Harvard Unit," looking up at his father. "I always remember he said to me, 'Well, son, I'm glad to see you
made it, old boy.'"

His father introduced him to two doctors, Harvey Cushing and
Richard Cabot. At the time "those names meant no more to me than they
would have if he said Smith and Jones. Now when I look back on it I
nearly die."[7] Cushing and Cabot, after all, were colleagues of Sir William
Osler, an association which in Scarlett's eyes almost amounted to medical
canonization.

That last battle remained in Earle Scarlett's mind, but so did the period
of recuperation which followed. He was profoundly impressed by the
treatment he received at the hands of the doctors and nurses in both the
field hospitals and British medical centres. "If his love of the humanities
was formed before the war, the latter sharpened his sense of human responsibility and he decided to study medicine as offering more stability,
usefulness and sense of accomplishment than the professor's life he considered," his daughter Katherine says.

He also liked to believe, according to Dr. Gardner, that while in a Manchester hospital he actually saw Osler. "After finding out about Osler,
what he did and how he looked," he once told his friend, "I'm sure I saw

him in a British hospital." Whether that was just wishful thinking is impossible to say, although Gardner points out that the timing was right. It was about the time that Osler's son was killed, and the great physician was in Britain while Scarlett was in hospital.

Scarlett, like many veterans, did not talk at length about the fighting, but he did like to talk about wartime medicine. Gardner, himself a battlefield surgeon in World War II, would sit in Scarlett's study, discussing medicine, music, or hiking in the Rockies. Scarlett would then turn to his friend and ask him, "Now, Smitty, tell me about Italy and tell me about your experiences."

"He was entranced by stories of wounded people," Gardner says, "and was forever comparing his experience from a soldier's point of view with medical thinking afterwards, comparing the two wars."

Some of Gardner's medical teachers had been medical officers during World War I, so he had some idea from them what Scarlett's experience of trench warfare had been like.

> Where they had ten casualties we had only one [in WW II]. And our techniques were very much more refined compared to those days. It was slapstick surgery. It had to be, because instead of twenty-five people to operate on they had two hundred. Then there was the poor equipment and so much in the way of infection of all sorts.

The problems of wound infection may have come to Gardner's mind because his mobile surgical unit in Italy received one of the first doses of the new antibiotic. The reason was that General Montgomery's headquarters happened to be close by. Montgomery eventually moved on, unharmed, and Gardner used the penicillin on a private with an infected wound with a result that seemed miraculous at the time.

Scarlett's convalescence went well, and while his wounds repaired themselves, the war came to an end. After he was released from hospital, he had a chance to tour both England and Scotland. One place fascinated him – Oxford. The idea of doing postgraduate work among the dreaming spires took hold, and he applied to enter the university. He also toured London, patronizing Ye Cheshire Cheese, an ancient pub a few steps off Fleet Street and near Dr. Johnson's house on Gough Street. To Scarlett this was a pilgrimage, since the venerable "Cheese" had been a meeting spot for Boswell, Johnson and other literary figures of the day.

Meanwhile, the letters between him and Jean continued. Finally, they came to an understanding that, on his return, they would become officially engaged. For a short time, however, it appeared that this reunion might be delayed for years, for in January he wrote to Jean to say that in hopes of being approved for Oxford he had taken his name off the list of men to be returned to Canada. A few days later, however, he cancelled his Oxford application. There was simply no way he could afford such a dream. In later years he returned to Oxford many times – once as an honoured guest in his capacity as chancellor of the University of Alberta.

Scarlett's diaries – and his reminiscences – do not reveal a man who revelled in martial glory. Like many of his generation, he saw his part in the war as a job to be done, and to be done as well as possible. Something of this showed through in a letter he wrote his bride-to-be after the conscription crisis which shook Canada. The Canadian government's reaction to the urgent appeals for reinforcements on the Western Front was to introduce conscription. Opposition to this, especially in French Canada, where there were anti-recruiting riots, led to a strange compromise. Conscription would go into effect, but only those conscripts who volunteered for overseas service would be sent overseas.

Jean's patriotism may have been piqued about this, for it's clear she asked Scarlett what he thought of this escape route (though her letter has been lost). "We don't give a rip about how many they let out," he wrote back. "It is better for the bunch out here to carry through as long as they can than turn a bunch of indifferent beggers out on us."

His wartime experiences in England allowed him to broaden his literary horizons and to reinforce his anglophilia. As he reflected upon the horrors of trench warfare, like many "old sweats" he could take pride in what Canadian soldiers had accomplished under the most inhuman conditions, but he could only hope that such a holocaust would never happen again.

Visiting England in 1938 on the eve of another war, he wrote in his travel diary: "My impressions of England are hardening. They show through three main currents: respect for the pageantry and the glory of the England of history, resentment at the enormous disparity in wealth and social position, and a half angry horror at the shadow of war and the way in which it is accepted as part of the scheme of things."[8]

Chapter Three

In his eyes she eclipses and predominates
the whole of her sex . . .
— "A Scandal in Bohemia"

Earle Scarlett took his demobilization in Toronto so that he could see his intended, and then, still undecided about which career he should undertake, he began the long journey back to Winnipeg. The Scarletts' war hero was happy to see his family again but unhappy with what he found in the Manitoba capital.

Writing in the 1979–80 yearbook of the E.P. Scarlett High School, he recalled 1919 not as the year of his homecoming but the year of "the strike": the Winnipeg General Strike. Having declared a general strike, workers were protesting in the streets. With anti-union sentiment and fear of communism running high among the authorities, the Riot Act was read. Some returned soldiers were among the 30,000 estimated strikers, but others, Scarlett among them, were drafted to help the Winnipeg police and the Royal Canadian Mounted Police (RCMP) disperse the protesters and control the riots. In doing so, Mounties fired into the crowd.

Scarlett told students at "his" school that he was slightly wounded in one scuffle with the rioters. In any case, as he wrote his fiancée, he deplored the actions of the strikers. The law had to be obeyed lest chaos reign. He held the same conviction years later when Saskatchewan doctors went on strike to protest the introduction of government medical insurance.

His father by this time was no longer an itinerant preacher but a well-established figure in Winnipeg, a noted minister and a man with an excellent war record. His son found him a valuable source of contacts. Scarlett had been thinking seriously about a medical career – he could not get the memory of that field hospital out of his mind – but before making a decision he decided to see what other careers might strike his fancy. The best way to investigate various professions, he reasoned, was to ask those already in them. Through his father he had met many of Winnipeg's leading professionals, and he made a list of between twelve and twenty men, many of them elders in his father's church. He then interviewed them, "pointing out that I was anxious to know something about why they had chosen their field of work." The results of this mini-survey, he told Dr. Roland, were "astounding."

> I found that over 90% . . . of these people, facing a returned soldier whom they felt, well, he's looked everything in the face, we're not going to keep anything from him . . . were sorry they had gone into their particular professions. I've never forgotten that. So that gave me the lead. So I looked around to see who were the contented ones. And do you know who headed the list? The two doctors I had interviewed. Not only that . . . but I found in talking to these other men in business . . . they said "you know we've had a successful career . . . but you know, oh I wish I had been a doctor because you have so much more satisfaction out of your work, you are helping people." Now that's not sentiment. Those men were talking to me out of their souls, and it's one of the greatest tributes to our profession that I know of.[1]

He told this story to many fledgling doctors, "hoping they wouldn't think I was a sentimental old fool."[2]

The survey fixed his mind on medicine. The place to study, obviously, was the University of Toronto. His father and grandfather had attended Victoria College, which had long since moved from Cobourg to Toronto, becoming part of the University of Toronto, and his fiancée was also a graduate from Victoria College. His problem was lack of funds. He was broke and he knew his father's views about grown men sponging off their parents.

Having already sought the advice of his father's friends, he thought he might see if a friend's father, an influential Winnipeg mason, might have some idea of how he could keep himself in funds. This man arranged an interview for him with H.S. Matthews, then general superintendent of dining and sleeping cars for the CPR, who had his office in Winnipeg. Matthews gave Scarlett a job as conductor, a prize for a student since his CPR pay would be augmented by tips, especially from rich Americans on vacation. He was given a good run, although a long one, on a train that began its journey in Chicago, ran over tracks owned by a U.S. subsidiary of the CPR to Portal, Saskatchewan, and ended its trip in Vancouver.

This particular job proved to be of immense importance to the fledgling physician. Scarlett's part of this run was from Portal to Banff. His stopovers resulted in a lifelong love affair with the town of Banff, the village of Lake Louise thirty-five miles to the west, and the mountains which dominated both town and railroad. Except for the years he spent as an intern and resident, Banff and Lake Louise would figure prominently in the rest of his life.

Back in Toronto, he settled into a rooming house run by a Scotswoman who had as a child worked underground in a coal mine and had later fled to more amiable surroundings in Canada. Here Scarlett ran into one of his first "celebrities," and he loved to recall the story of that meeting to test the knowledge of young doctors. In the 1960s and 1970s, as a retired patriarch of the medical clan, he would often receive a young physician seeking his medical blessing. He told me of one nervous young man brought to his study who, wanting to break the ice, asked him about his medical days. "Well," Scarlett said, "I told him that one day I came back to the rooming house after classes and found my landlady sitting at the kitchen table. On the table was a half-empty bottle of scotch and sitting on the other side was a shirtless man wearing a kilt. 'Dr. Scarlett,' she said – she always called me Dr. Scarlett – 'I'd like to introduce you to a childhood friend of mine, Sir Harry Lauder.'"

Dr. Scarlett, who expected a physician to be a man of culture and assumed any man of culture would recognize the name and be impressed, waited for a sign of admiration from the young doctor. The physician simply looked blank.

"Who?" he said.

"You'll never be a doctor," Scarlett thundered, and the interview was over.

The chastened physician probably spent some time leafing through a book of medical notables to identify this Scottish physician. Lauder, of course, was the famous Scottish music-hall singer who had died in 1950, long before this episode took place, and who was knighted in 1919 for his work in raising money to support the war effort. To Scarlett such a man should not be forgotten, especially by a doctor.

Scarlett entered medical school with other veterans, many of whom would make names for themselves in the profession. Like him, many were hard up. A fellow medical student and friend was Robert McClure, who became a medical missionary and the first lay moderator of the United Church. McClure, Scarlett remembered, lived in quarters at Knox College. "He enhanced his income by cutting the hair of fellow students at fifteen cents per head. He learned this art during two summers when he served as intern in the Mental Hospital at Cobourg where he cut the hair of male patients."

A professor of note was Fred Banting – or "Specs," as he was called by students because of his prominent glasses. In Scarlett's second year, Banting was his demonstrator in physiology. "It is now history that in the hot summer of 1921 he worked on his diabetic dogs in search of a secretion to control the disease, his lab being not much bigger than a closet. He was assisted by a student in the year behind us, named Charlie Best. He once told us that his research budget for that summer had been $100. And insulin was the result."

The teachers Scarlett remembers best taught him more than medicine. One such was Dr. Jabez Elliott, a tuberculosis and lung specialist, who did his best to interest students in the history of medicine. Whatever success Elliott had with the other students, he certainly captured Scarlett's interest.

Another was his anatomy professor, J. Playfair McMurrich. McMurrich's lesson was in how to live, something else Scarlett would not forget.

> We were chatting one day in his room, it was the old anatomy building which had the smell of corpses of 60 years. Boy, it was strong in the nostrils! We were sitting in his little room, study, all kinds of framed prints around the walls [and] . . . I'll never forget, these were his words, he said, "You know, Scarlett, the way to be happy in this world is to make use of it, to make what you love your avocation, and make your vocation the means by which you make your bread and butter. Now," he said, "just to show you that I'm not talking idle nonsense to you, anatomy is my vocation." I said, "Dr. McMurrich, you're a world-known anatomist. You've published

an anatomy that's in every medical man's library, and yet you tell me that's second." I said, "Would you tell me what your avocation is." "Yes," he said, "with enthusiasm. Leonardo da Vinci's etchings." And if you go and look it up you'll find that the definitive great study and so on of Leonardo's etchings is by J. Playfair McMurrich.[3]

There's an echo of McMurrich's words in Scarlett's advice to high school students that was printed in their 1977 yearbook, although Scarlett used the term "hobby" rather than "avocation." "Settle on an agreeable hobby as early as you can," he wrote, "something you can love and in which you can lose yourself when off parade. This is one of the main sources of happiness which cannot be purchased by any cheap fare to pleasure domes."

Scarlett wasn't the sort of student who never got into trouble. In fact, he was suspended twice, for two weeks on each occasion. One expulsion came after Scarlett, with Murray Meekison and a number of other students, became involved in a row with Dr. Beverley Hannah at "Sick Kids." Earle never said what his specific crime was, but suffice it to say that returning veterans were not always compliant students. Nevertheless, his talents were such that his fellow students at one point elected him president for the year.

Scarlett's avocation was writing, and his talents in that respect also seem to have been recognized by the students. One of his proudest moments at university came when a committee of students approached him to edit a medical journal that they envisaged as a tribute to their class. The funds would come from the class. That the class could afford such a venture may be explained by their status, for these were no callow youths. Dr. Scarlett described himself and his fellow students as "a tough bunch," all veterans who had seen service in France and who were more mature than the average medical student; as adult students, perhaps they had had better things to spend their money on than social events. In any case, they accumulated $1,200 in dues, an enormous sum for the period.

Scarlett, twenty-eight years old by this time, thought a journal would be a splendid idea and recruited Murray Meekison and Robert Stone as his assistants, and they founded the *University of Toronto Medical Journal*.

Instead of gaining him kudos, as he told Charles Roland, it got him into trouble with the professor of medicine, Dr. Duncan Graham, who was not one of Scarlett's favourite people.

He wanted to see me so I turned up and he was in a foul mood. Maybe that was his usual mood although Ray Farquharson, who happened to be one of my dearest friends on this planet . . . told me afterwards "you've got him all wrong. You didn't get him sized up at all. He's a very decent chap at bottom." "Well," I told him, "I'm glad to hear it because he sure treated me badly." And he was foul that afternoon. And afterwards I discovered why. He said to me, "I see, Scarlett, you and some of the boys in your class have started a medical journal around here." And I said, "Yes Sir." He said, "That's no good. We can't have that sort of thing going on around here. Whose permission did you get?" I said, "We got the permission of the Dean and the Faculty Committee on Publications." So he couldn't dispute that. "Well," he said, "how you gonna pay for it?" I said, "Don't worry, we've got that covered." Anyway, we went on and on and he was savagely critical, he picked the thing up and picked holes in it and finally he said, "I'm going to make a prophecy (I'll never forget this), Scarlett. This thing won't finish out a year. Students can't edit a medical journal." So I said, "Thank you very much."[4]

Graham, after predicting the journal would go broke, was evidently astounded when he discovered how much money the class had kicked in for the journal. According to Scarlett, Ray Farquharson, a friend and associate of the professor's, later explained Graham's attitude. Graham and another well-known member of the staff had attempted to start a journal a few years before and it had failed.

The friction may have also been a simple clash of personalities. Both were headstrong Scots and veterans of bloody campaigns. Graham, as a medical officer, had served under canvas at the base hospital during the miserable and costly debacle in Salonika.

Each summer, Scarlett refilled his coffers by going back to work on the CPR. Since the train on which he worked originated in Chicago, it was often full of Americans, whose manners he disliked. He did, however, like their money. At the time the U.S. dollar was worth $1.21 Canadian, and when Americans paid for everything at par, he was able to pocket the difference.

During his first year of studies in Toronto, the one person he most wanted to see was the person he saw the least. Jean Odell graduated from Victoria College in 1919 with high honours and took up Scarlett's old vocation, teaching. She was hired, not by an Eastern school division, but one in the West. She ended up in a one-room schoolhouse in Dee Valley, Saskatchewan. Their love affair had to be carried on mainly through the royal mail – a typically Victorian method, perhaps, but not exactly satisfying.

At the end of her first year as a teacher, they were able to spend part of July together in Calgary and Banff. Although a chaperone may have been present, or somewhere in the vicinity, it was a romantic time for both of them, one Earle referred to as their Banff "honeymoon." They dined at the Palliser, the CPR's luxurious railway hotel in Calgary, drove to Banff and took a cottage in the pines.

His letters after this were rife with lush, more or less Victorian romanticism. "We have found the key, as Stevenson puts it, to the city of love," he wrote her in August.[5] "Last night I climbed Tunnel Mountain to make a pilgrimage to 'our tree.'"

After a year on the bald prairie, she moved to another school, this time in Port Hope, Ontario, but eventually she returned to Toronto. This forced Scarlett to learn yet another art.

His college yearbook may have called him a "lady-killer," but in a private manuscript written after his wife's death, he notes that he had hardly had a date before the war, much less learned to dance. He soon discovered that if his romance was to progress smoothly, he would have to take this up. His room-mate, another medical student, was also unskilled on the dance floor, so the two of them registered at Mosher's Dancing Academy and took lessons.

Like other penny-pinching students, he ate some meals in the dining room at Hart House but occasionally would dine out with his friends at a café near the corner of Spadina and College known as "The Greasy Spoon." Dinner cost twenty-five cents.

In 1922 he returned to the CPR but for the last time. By the summer of 1923, he was learning the practical side of medicine, not to mention golf, as an intern at the Portage la Prairie Medical Clinic. While today's interns are noted for having to work long hours for modest pay, Scarlett had to pay for the privilege of interning. Almost fifty years later, when as a member of the board of Calgary's Foothills Hospital he found himself involved in labour negotiations, that experience came to mind. Not, he admitted, that it did him much good.He wrote to his friend William Bean, M.D.:

At the moment we are engaged in close combat with the residents and interns who, in the matter of salary, have developed delusions of grandeur. I feel my age when I ponder their demands – no use my telling them that in my first intern year I paid $25 a month for the privilege of interning – no such thing as a salary. I also feel sad when I think these same fellows, when they go out into practice, will probably charge patients on the same bloated scale to which they have been accustomed. There is a poor chance of St. Francis of Assisi being adopted as one of the patron saints of contemporary medicine. At the moment I see no signs of any humility.[6]

It is true enough that Scarlett was not overly concerned with money and that sermons he had doubtless heard about the love of money being the root of all evil had left an impression on him. This being so, one of the ironies of his life is that the man he referred to as "the patron saint of my profession of medicine," Sir William Osler, had no compunctions about charging large fees. He died, in England, a millionaire.

In March 1923 Scarlett had to return to Winnipeg when his father died at the age of fifty-seven. His obituary in a Winnipeg paper was, as befitted his position, long and fulsome, praising him as "a preacher of the uncompromising type . . . vigorous in speech and attacked whatever he thought was opposed to the cause of religion."

Upon graduation in 1924 and after arranging for a residency at the Henry Ford Hospital in Detroit, Scarlet married Jean Odell. What they called their "Olympian honeymoon" was spent in a cottage beside Lake Muskoka, and though their diaries show a certain division of labour, Jean was obviously overjoyed to have the chance to keep house for her husband. Romantics at heart, the two pasted notes, poems and keepsakes in "Our Memory Book 1919–1926," which was "dedicated to the God of Happiness, the patron saint of December and the patron saint of Love."[7] The honeymoon over, Earle left Jean in Cobourg and went to Detroit to find housing suitable for a new physician with a tiny purse. His bride, her teaching career exchanged for that of a housewife, joined him in a small basement apartment on Seward Avenue near the hospital.

Scarlett soon discovered that his wife had one quality he lacked: she knew how to handle money. From the very beginning of their marriage, Jean was the family treasurer. "She was a good one, too," he admitted, noting she had been brought up by a Scots aunt and had to scrimp during her college days. She had in fact hoarded her money from the time she had first taught school, and her savings, in addition to money from various scholarships Earle Scarlett had earned upon graduation, paid for their honeymoon and a few necessities. Now, as a resident, he would be paid $150 per month, four dollars more than Jean's last teaching assignment had brought in.

"I will never forget our first evening in our new home," he wrote in an unpublished series of notes. "Jeanie was finishing her accounts and finally closed her little book. 'Well,' she announced, 'we are solvent. We have a 65 cent credit balance.'

"I took up my duties the next day in the happy assurance of our financial state of health. So you may say, we launched ourselves into married life on 65 cents!"

Not only was he financially solvent, but he had dropped into a very good position. While many of the residents at Henry Ford were on rotation, Scarlett and a few others managed to hook themselves into a gastrointestinal program in which 95 percent of their time was spent learning the practical side of patient care. During the afternoon he was free to roam the wards and study interesting cases without a corresponding obligation to provide service. "It is amazing that the hospital would allow such a straight instructional course for us without asking for more in return," he wrote Jean in September of 1926 while she was visiting relatives in Canada.

During this residency he turned out the first of his medical articles; but Henry Ford hospital was beginning to pall. Having picked up an interest in coronary artery disease, he was happy to respond to a call from Prof. Fred Smith, a pioneer in the field at the University of Iowa.

By this time he had also developed a fascination with Osler which, Bill Bean believes, may also have had something to do with his choice of Iowa, a state he knew nothing about other than its reputation for growing corn. Bean, in an essay on Scarlett, wrote that the presence of "Campbell Howard, son of Osler's revered mentor Palmer Howard, no doubt attracted him to Iowa City." Unfortunately, he added, "Howard was leaving to take the chair of medicine at McGill University in Montreal when Scarlett arrived."

When Scarlett arrived in Iowa in 1927 he found he had a title: "associate in medicine." He never did figure out just what the title was supposed to mean, but he did enjoy the money, "a hell of a big salary" of $175 per month. He sometimes was able to enjoy a second title, since the university listed him as a "Fellow" in the medical faculty.

His three years in Iowa were useful in honing his skills in both his vocation and avocation. He delved further into internal medicine and coronary heart disease and mastered a new and not always accepted machine called the electrocardiograph. A few years later he would introduce the electrocardiograph to Calgary, only to discover the medical community was not impressed.

His colleagues at Iowa were impressed, however, with his clinical work, and they were aware of his interest in literature. This led to his first medico-historical paper, a long essay on the history of the University of Iowa Medical School. The fifty-page, unsigned article was printed as part of a special booklet produced to mark the opening in 1928 of a new general hospital and medical school which had been financed by the Rockefeller Foundation.

He also began to publish in academic journals. These early articles focused on clinical medicine and medical history. The first article, published in the *American Journal of Medical Science* and still cited in the 1980s, was on the significance of severe anemia in chronic nephritis. The second, on the history of plague, appeared in the *Medical Journal and Record*. These were followed by another paper on nephritis, in the *Journal of the Iowa State Medical Society*, and an essay on Jean Paul Marat, a physician and notable figure in the French Revolution, in the *Annals of Medical History*.

Meanwhile, his family grew with his reputation; in 1928, Jean and Earle had a son, Robert Michael.

As the months went by, he found work in Iowa less and less to his liking. He began to think about returning to Canada. According to A.W. Rasporich, this was spurred on partly by "cultural nationalism and nostalgia,"[8] and partly because of an ill-defined anti-American feeling. These may have been factors, but Scarlett told Charles Roland that he found that clinical standards were falling at the university hospital and materialism on the campus was growing. "A man teaching at the university, as far as I can see, was rated by the car that he drove."[9]

One day, he came to a sudden decision.

> I came home and I said to my dear wife, "Jeannie," I said, "we're packing up and getting out of here and we're going back to Canada." She said, "Why?" I said, "I can't stand this choking materialism any longer. Something has happened. The spark has gone out of this so-called institution of learning. We're going back to Canada if I have to dig sewers in Medicine Hat."[10]

This was a fine idea, but at this point Scarlett was unsure where to go. His first thought, naturally enough, was Toronto, where he had experience, knew some of the medical community and might be welcomed as an internist. A letter to his friend Ray Farquharson, who had stayed in Toronto and was advancing up the medical ladder at Toronto General Hospital, brought a reply stating that he was sure Scarlett would do well in Toronto and to return post-haste. Farquharson was too optimistic. Fatted calves, as it turned out, were not in the offing in Toronto.

With the move in mind, Scarlett next wrote to his old nemesis, Dr. Graham, who was chief of medicine. The reply he received was cool. The medical powers-that-be would welcome Scarlett back to Toronto and would give him an appointment on Toronto General's staff. They would even give him a private office and a small honorarium for doing demonstration work in the outpatient department. But there was a catch. All this hinged on his appearing, in person, to apologize to Dr. Duncan Graham for having had the audacity to move to the States for his postgraduate work. As Scarlett told Smitty Gardner years later, there was no way he would grovel before anyone, much less someone he did not respect, so he fired back a letter refusing the offer.

He wrote to his mother about his plight, and she replied with a different suggestion. Instead of writing to a doctor, she said, write to a lawyer, in particular one R.B. Bennett, who knew of the family. Bennett at that time was Conservative MP for Calgary West *and* Canada's prime minister. Scarlett wasn't the only ex-patriate Canadian veteran to write to the prime minister for the favour of his advice.

Bennett replied to the young doctor suggesting he write to his friend and long-time political ally Dr. G.D. Stanley of the Calgary Associate Clinic. Calgary appealed to both Jean and Earle, bringing back memories of their first "honeymoon." Earle sent off a letter and Stanley replied with a job offer.

Unsure what awaited him in Calgary and not knowing where he would stay, Scarlett dropped Jean and Robert in Winnipeg with his mother, hopped on a west-bound train, this time as a paying passenger, and headed for the city which would be home for the rest of his life.

Chapter Four

Here is a gentleman of a medical type,
but with the air of a military man.
— *A Study in Scarlet*

CALGARY HAD GROWN SINCE E.P. Scarlett, CPR conductor, had known it. In 1921 it had still been a pioneer town full of characters: *Calgary Eye Opener* editor and publisher Bob Edwards, lawyer Paddy Nolan, Senator Patrick Burns, and assorted ranchers and Indians. Now it was Alberta's largest city, a title it would lose during the Great Depression to Edmonton. It was a bustling and friendly place of 83,000 people and two relatively large hospitals.

But Scarlett had also changed. He had a wife and a two-year-old baby boy, Robert Michael. And he was now a medical specialist.

He arrived on familiar territory, the CPR station, and was met by the prime minister's friend, Dr. Stanley, who immediately took him to what would become even more familiar ground, the Holy Cross Hospital, to introduce him to his new colleagues. They were, as it happened, in the operating room, and Scarlett described the scene in 1954, when he unveiled Stanley's portrait at the Associate Clinic.

First of all he introduced me to a pair of eyes above a surgeon's mask. Those eyes turned out to belong to Dr. Macnab. The setting as I saw it on that occasion was typical. Doctor Macnab was as always operating intently as if his own life depended on the issue; Doctor Murray sitting in relaxed fashion at the head of the table was giving the anaesthetic; Doctor Aikenhead was assisting, moving about the scene, saying little and doing his job; Doctor Lincoln was in the dressing room discussing in a fine flow of conversation some topic of world affairs. Doctor Stanley was the host and spokesman for the group. It was a characteristic setting that I was to become so familiar with in succeeding years.[1]

Scarlett, who had already worked in a clinic setting as an intern in Manitoba, was obviously impressed with the collegiate atmosphere, but he may not have realized just what he was getting into. "There are probably no enmities that go more deeply than academic and medical ones, and the currents and countercurrents that flowed around the roots of the medical profession in this town were really something," Scarlett remembered.[2]

There existed a deeply rooted prejudice against the idea of a clinic, and in 1922, when five doctors – J. Scovil Murray, A.E. Aikenhead, W.A. Lincoln, G.D. Stanley and D.S. Macnab – decided to form one, they didn't dare call themselves a clinic. Instead, they rented offices together in a sandstone building in downtown Calgary and called themselves "Associated Physicians and Surgeons." By the time 1928 rolled around, the group was firmly established and changed its name to Calgary Associate Clinic. This, however, reflected the group's lack of fear of their own colleagues, who for the most part had been converted to the idea of a clinic, to which in 1930 many physicians were still opposed.

Nor was the prevailing attitude about to change overnight.

Donald McNeil, M.D., an eminent city doctor who came to Calgary to join the clinic found the same emotions in 1940.

To many people in Calgary it was really a wonderful place. It had a great reputation. To other people whose doctors were not in the clinic it might produce a little bit of a smile or even disdain. They would say that "it was a machine" or "it was a mill that you got into but you couldn't get out." It was criticized like this by the other physicians. This is not without exception because there were men in the city, very good men who had considerable respect for particular members or particular men in the clinic. But to a certain number of physicians we were practically evil. We weren't of the high ethical standing that they thought they were. "If you sent a patient to that damn clinic, you would never see the patient again," they would say. The patient would be lost and wouldn't come back to you. This is what they condemned us for more than anything. This was to some extent true.[3]

As for Macnab, McNeil said, he lacked respect for a number of city doc-
tors and had no compunction about telling his patients what he thought.
"Doctors then didn't hesitate to speak badly of other doctors to their patients.
'You know he's an old butcher,' they might say."[4]

Meanwhile, Scarlett, who loved discovering the derivation of words, was
amused when he came across the possible root of the word "clinic."

> All our readers who share allegiance to that company of the medical world
> known as "clinics" will feel the same lively interest that I had in the follow-
> ing item. It appears in a book published in 1730 at the "Ship" in Paternoster
> Row, London, for J. Osborn and T. Longman. The author is one John Quincy,
> M.D. The title of the book reads: *The Lexicon Physico-Medium or A New Medi-
> cal Dictionary* "collected from the most eminent authors and particularly
> those who have written on Mechanical Principles." In its pages I find the
> following definition of clinic: "Clinick – generally used to signify a quack; or
> an Empirical Nurse who pretends to have learned the Arts of curing disease
> from attendance upon the sick."[5]

Macnab, who died in 1951, was the guiding spirit behind the clinic. A
Nova Scotian Scot with a penchant for cursing and who, outwardly at least,
gave an impression of "bluster and noise," he was nonetheless a canny man.
With his chief associate Stanley, he studied the operation of other clinics to
discover why some succeeded and some failed. One result of this investigation was the hiring of the clinic's first business manager. But business sense
alone did not make for a successful clinic in Macnab's terms. That would
only come through the members' mutual respect and loyalty, he told Stanley.
"Loyalty was his prime demand and his personal loyalty to each individual
associate was unbounded."[6]

To tie the associates together and to keep standards high, Macnab insti-
tuted weekly clinical luncheon meetings, a tradition which continued well
past his death. To Scarlett, who also valued loyalty, Macnab became a figure
to emulate second only to Osler.

As for Stanley, who shared Scarlett's love for medical history, he became
an important figure in the internist's life for the next thirty years. Stanley,
who had come to Calgary in 1918 after practising for seventeen years in the
ranching community of High River, had built up a city practice which included Bob Edwards and a CPR lawyer named R.B. Bennett. According to
Scarlett, Stanley's main job in his first years in Calgary was to keep Edwards

sober, a task Stanley himself called "a continuous and helpless professional struggle."[7]

There were other clinics in the city, the McEachern Clinic and Mackidd Clinic, both named after their founders, but their doctors tended to use the General Hospital on the north side of the Bow River, which divides Calgary. (The General Hospital, in a manner of speaking, doesn't exist any more. In a move evidently designed to confuse Calgarians, the Calgary General Hospital is now two hospitals several miles apart. The old General is officially the Bow Valley Centre.)

Doctors with the Macnab Clinic, as it was sometimes called, did their work at the Holy Cross, a smaller hospital run by the Grey Nuns on the south side of the city. Scarlett lost no time in applying for privileges at the hospital, submitting the usual papers and letters of reference. If the practice had been to deny an applicant copies of his letters of reference, Scarlett might have never seen the comments of H.S. Houghton, M.D., dean of medicine at the State University of Iowa. Scarlett, Houghton wrote to the Sister Superior on 27 September, 1930, "was highly successful as a teacher" and had "contributed in an important way to the teaching, research, and general clinical activities of the institution."

Scarlett was given admitting privileges on 16 December 1930, and on 1 January 1932 became an active member of the hospital staff. It was a connection he relished and which remained until 1954, when pressure from his other duties at the clinic and as the University of Alberta's chancellor forced him to switch his status from "active" to "courtesy." In February 1956, M.D. Mitchell, M.D., on behalf of the executive committee of the staff, and Sister Rose Leteller, the administrator, asked Scarlett to accept honorary membership on the staff. "May God bless and guide you always in your noble endeavors to instill in others your high ideas of a sound and practical philosophy of education," the sister wrote.

The year 1932 was marked by his obtaining the fellowship of the Royal College of Physicians and Surgeons of Canada. He became a fellow of the American College of Physicians in 1946.

While the Associate Clinic doctors occasionally worked at the Calgary General Hospital, they didn't feel comfortable there, perhaps, McNeil thought, because they were not treated as royally as they were at the Holy Cross. Another reason may have been the ill feeling between the principals of the different clinics; Dr. Mackidd and Dr. Macnab would cut each other dead if they happened to meet.

The reason for the royal treatment accorded Macnab's doctors at the Holy Cross was simple. The Associate Clinic's practice of pouring its patients into the hospital was a big factor in keeping it financially healthy, as the sisters well knew. What the clinic doctors wanted, one nurse remembers, the clinic doctors got.

Besides his being a clinic doctor, there was a second reason Scarlett was not a universal hit with his fellows: he was brought in as a specialist. He was not the first. The clinic had brought in a pediatrician, Dr. Price, in 1928, and Macnab had been certified as a specialist in general surgery the same year; but specialists in internal medicine were rare birds and distrusted. According to Scarlett, some Calgary doctors "sent him to Coventry"; he would walk into a room filled with other physicians and they simply wouldn't talk to him. In his first five years in Calgary he was consulted only twice by doctors outside the clinic. The arrival the same year of a specialist in obstetrics, Dr. Fisher, did even less to endear the clinic to other doctors.

The reason, Scarlett believed, was that almost all of the physicians in the city had followed a similar career path. They had graduated from medical school, set up a general practice in some rural area, and had then moved to the city. Scarlett had never intended to set up his own private consulting practice. Nor had he been in general practice, and this, he felt, caused a good deal of resentment. In centres with a university and a medical school, minds were more open, but Calgary's medical school was still more than forty years in the future. Scarlett's disgust was especially keen because of the reception given the electrocardiograph he had brought with him – the first one in Calgary, he believed. He expected this modern device would bring him kudos. Instead it was regarded as "a magician's fake gadget," with the result that the only electrocardiograms he did were for his own patients and those of his associates. History has to some extent made up this injustice. Harold N. Segall, who trained in cardiology in the 1920s and who has written a history of cardiology in Canada, has no hesitation asserting that Scarlett and his machine "represents the introduction of twentieth-century cardiology into the Calgary region of Alberta."[8]

The clinic had opened in rented quarters, but by the time Scarlett joined the group it had moved to a new building in the heart of Calgary, one which had its own lab, run by Fred Langston. It remained there, until the 1970s.

Soon after Scarlett's arrival, the Depression hit with a vengeance. But the clinic flourished. Money was scarce and Scarlett's salary was sometimes only $100 per month. Even so, he was luckier than many, and he knew it.

Once during the mid-thirties he visited an elderly patient. The woman was dying and Scarlett found her spending her final days in a cold and tiny rented room, with little to eat and no one to look after her. When he discovered other elderly patients in the same plight, he decided to act. With help from some of his friends, he managed to find a small house near the Holy Cross and turned it into a home for the aged, the first, Scarlett believed, in the Calgary area. He tried to keep his own home and his practice separate and he said little about this establishment to family members, although he once showed the house to his daughter Kay. The house meant a good deal to him. "Father would drop in regularly to visit his 'old ladies,'" Betty remembers. "They adored him."

Scarlett became a familiar figure at the Holy Cross, and through his lectures to student nurses he eventually lived up to Houghton's compliments on his teaching. These lectures may have been the only occasions where he did not mix medicine with literature or pepper his lectures with classical quotations. Those he taught do remember, however, that he could not resist bringing in Osler and his work. While he was a tough teacher who demanded strict attention from his students – and from nurses on the wards – he was respected and liked.

Dr. McNeil, after his return from World War II, was given a staff position at the Holy Cross. As part of his duties he organized lectures for the nursing students. He chose Scarlett, who had given up his Holy Cross lectures some years earlier, to deliver the introductory and other lectures. "The nurses just loved this," McNeil said. "When he retired I asked him for a copy of his notes and he said, 'Sorry, I don't keep any notes. I just talk right off the top of my head.'"[9]

Scarlett had the deepest respect for the nursing profession. As the chairman of a commission on nursing education, he would eventually come into official conflict with it, but nurses who remember him do so with affection. In the 1960s and 1970s, long after he had retired and when his weak heart forced him to stay close to his house, nurses from the hospital would often drop by just to visit.

Dr. McNeil, fresh out of school, worked under Scarlett both at the clinic and at the hospital. One of his duties was to work the clinic's antiquated X-ray machine, a device which sometimes shot sparks across the room and which Scarlett feared would some day electrocute either a patient or the operator. McNeil's X-ray training was far from extensive; Scarlett simply took him into the X-ray room one morning and gave him two hours of instruction on that machine as well as on the clinic's fluoroscope. The instruction must have been reasonably good, for McNeil never suffered from the X-ray burns that affected many physicians at a time when the real dangers of X-ray radiation were not well understood.

Working under Scarlett, Dr. McNeil discovered one reality of practice: the long hours. "I was in that clinic and could be reached at any time of the day or night. I was given Thursday afternoons off, but I was told that didn't include the evening," he recalls.[10] And if he followed his mentor's example, it seemed the pressure would never let up. Scarlett appeared at the Holy Cross switchboard promptly at 8:30 a.m., having made a house call or two. He would then set off on his rounds, Dr. McNeil respectfully behind him. He would see his own patients if he had any, or other clinic doctors may have asked him for a consultation on a case.

He always saw a fair number of new patients every morning. He didn't spend a great deal of time with them. I went along with him as well, as my time would permit, and indeed I did learn from him. He was a good teacher. His knowledge of medicine was broad.

At 10:30 he would leave the hospital and perhaps he would have a house call. Many of the very senior people in this city expected him to look after them, like the judges, the lawyers, the senior doctors and businessmen. He did look after and advise these people and carried on their treatment, whatever it was, not just as a consultant but as a practicing physician. He would be at the office by, say, 11 a.m. He would have cardiograms to read. He brought in the first cardiograph instrument into the city . . . a beautiful machine: the record was photographed on a film . . . He would then see one, two, maybe three people before 12. He would be promptly finished at 12. He might then go home to lunch or he might be attending one of the clinic's luncheons.[11]

The clinic luncheons begun by Macnab had grown. By the 1940s the weekly affairs were held in one of the posh meeting rooms of the Palliser Hotel, with the clinic's president – Dr. Scarlett took over from Macnab in 1947 and held the position until 1956 – presiding at the head of the table. The clinic's specialists would give talks on their specialties, someone would have been delegated to review notable articles in the latest medical journals, and those returning from meetings or courses were expected to discuss what they had seen and heard. The Palliser provided a good spread, but the meeting would be dry; Dr. Macnab would not allow alcohol at a professional meeting and Scarlett agreed with this, although he served drinks at social gatherings in his home.

Promptly at two o'clock Scarlett would return to his office to see patients with appointments and any emergencies that turned up. He would also dictate letters to the clinic's secretarial staff, a privilege the clinic allowed him long after his active days were over and when he was, in a sense, senior member emeritus. Somewhere between half past five and six he would finally leave the office for his evening hospital rounds, followed by dinner and more house calls. He loved night work and would sit late in his study, writing letters or columns, or entertaining friends. As he grew older, he took Saturday afternoons off, then the whole of Saturday unless a patient needed him.

After watching Scarlett for years, McNeil had considerable respect for the older physician.

> He did provide service, he was well regarded and he was capable. He was well trained to begin with. Dr. Scarlett did belong to many associations. He had many activities, but he did regularly attend the very best of medical meetings over the continent. Earlier he travelled by train but he even attended more when he was able to fly. So he was well trained and generally, he brought to Calgary one or two new things every year. I can remember him returning from a conference and that was when we began to talk about anti-coagulants. That's when we began to use them.[12]

McNeil didn't think Scarlett was above criticism, however. Scarlett had a well-earned reputation as a diagnostician, but though McNeil respected his abilities he felt Scarlett spent too little time with patients. According to a nurse who worked with both of them, McNeil's opinion reflected his own way of working. Scarlett would pop into a room, write up his notes, and leave. McNeil, on the other hand, preferred to chat. "Dr. McNeil could spend an hour with a patient and get very little done," she said.

As Scarlett's outside commitments became heavier, the clinic's tradition of mutual respect and loyalty worked to Scarlett's advantage. Administrative work took him away from his patients to some degree, but there were also more meetings to attend and other demands on his time, especially after he was appointed chancellor of the University of Alberta in Edmonton in 1952. He was no stranger to that institution, having served on various boards and committees, more often literary than medical. His absences increased the workload of Dr. S.B. Thorson, Dr. Houghtling, Dr. Muldoon and others, but they accepted the extra burden with good grace.

As president of the clinic, Scarlett proved an able administrator and organizer, although it was probably just as well that the clinic had a good business department led by such people as Ed O'Connor. One of Scarlett's weaknesses was an inability to handle money or even, in some cases, to submit what colleagues considered to be a reasonable fee. Dr. McNeil remembers one of Scarlett's patients, "an extremely wealthy man," who was treated in the hospital for two weeks during a grave illness. Scarlett gave the man excellent care, handled constant phone calls from the family, and in the end submitted a bill for $35. To juniors at the clinic, the financial matters of the clinic tended to be something of a mystery, known only to the most senior partners.

Both Scarlett and Macnab left their mark on the clinic and on medicine in Calgary. One of Scarlett's innovations, and one which earned him medical friends throughout North America and abroad, was to take on two fellows each year, one from medicine and one from surgery. At the end of the year, he asked them how the fellowship could be improved and offered to help them with further training. At least ten of these fellows were sent on to the Mayo Clinic.

He also improved his own education whenever possible. In 1938, with his wife, he paid a working visit to England, which allowed him to revisit many of his favourite spots. He added to his skill by studying at the National Heart Hospital and the National Hospital for Nervous Diseases in London, and by acting as a visiting resident in Edinburgh.

While Scarlett practised as a physician, his other activities continued to grow. Calgary in the 1930s was still a cultural backwater set in the Canadian bible belt. Scarlett had to take on new challenges to keep his mind alive. In 1931, and off and on for the next twenty years, he taught sex education classes to boys at the Young Men's Christian Association (YMCA), an enlightened service which did not endear him to those who enthusiastically endorsed the precepts of Alberta's premier and fundamentalist preacher, "Bible Bill" Aberhart. The city police, in fact, received complaints about these classes, although no action was ever taken.

In 1932 he joined the board of the Calgary Symphony Orchestra, a connection he kept alive for thirty-six years. From 1934 to 1937 he was a member of the city library board, and from 1938 to 1940 he was chairman of the regional advisory board of the Canadian Broadcasting Corporation. It is difficult, in fact, to find a group with which he was not connected. Many wanted, if not his services, then at least his name. A partial list would encompass groups such as the English-Speaking Union, the Boy Scouts, the Salvation Army, the YMCA, the Canadian Institute for International Affairs, the Classical Club of Calgary and the Canadian College of Organists.

And, of course, he continued his writing, but now for Canadian journals, since he was back in Canada. Articles he wrote with Dr. Macnab for the *Canadian Medical Association Journal* covered such topics as the biliary system, the value and administration of glucose in surgical and medical conditions, phenobarbital poisoning, and peptic ulcer. The *Historical Bulletin* (see Chapter 5) gradually took up more of his writing time, but between 1932 and 1939 he published thirteen papers on medical subjects and five on history, most of them in the *Canadian Medical Association Journal.*

He loved the poetry of John Keats and joined both the local and the New York chapter of the Keats-Shelley Association. He also discovered the Baker Street Irregulars, that odd association of fans of the work of physician-author Sir Arthur Conan Doyle, alias John Hamish Watson, M.D., and the consulting detective he met at "Barts," Mr. Sherlock Holmes. He was delighted when Dr. Stanley, telling stories of his days in practice in High River, mentioned that "remittance men" (scions of British families sent to the Wild West and paid to stay there) would saddle up and ride the forty miles into town to pick up their copy of the *Strand* magazine so they could read the latest adventure of Holmes and Watson. Scarlett knew, either in person or by correspondence, most of the famous Irregulars. He told Stanley's story to one of the movement's founders, Christopher Morley, who was visiting Calgary. The story later turned up in one of Morley's last published works.

Soon after the outbreak of World War II, Scarlett added to his duties by helping set up the medical side of the Colonel Belcher Hospital. Although he was an internist, he was one of those charged with finding surgeons to look after wounded soldiers on their return to Canada. In 1947 he was named senior consultant in medicine at the Belcher, a position he maintained until 1958. Dr. Gardner, who was initially surgical consultant, then chief of surgery at the Belcher from 1945 to 1971, became one of his closest friends. He admitted Scarlett had his faults but he had high praise for his friend as both internist and teacher.

He had a very clinical approach. He had a beautiful mind, he was well schooled, well informed, so that there was no variation in his approach to a problem. And so in that respect he was a good internist. But he was so busy with so much that he could be a little behind the times – not clinically but technologically a little bit. Moreover he could be unaware of variations of various syndromes because of the lack of time to read. In order to be a really first-rate clinician you have to read constantly. There's a new syndrome every week, or a new approach to an old syndrome. In order to be really conversant in that area you've got to read what was printed today or last week. But when he gave a clinic, which he did every week at the Belcher, I used to sit in. I was fascinated with the way he did it. I wanted to learn, brush up after the war years, not only surgical but medical disease. He approached everything in a very learned way so that he was a good teacher. But he was considered a good internist and he was considered *the* cardiologist during those early postwar years.

One thing Dr. Gardner liked about his friend was Scarlett's delight at telling stories against himself, especially to prove a point.

One of his best friends came and said: "Look Earle, I've been told I'm due to have heart trouble. I want to see what you think." So he gave him an ECG, and checked him all over, and he was such a friend he saw him out of the office and right out the door to the sidewalk. As he bid him goodbye, slapped him on the shoulder and said: "You're fine, stop worrying." The man stopped worrying right enough. He dropped dead of a heart attack at Scarlett's feet.

"He used that as an example of how little we knew," Dr. Gardner said. After Scarlett's death, Gardner wrote an appreciation which was presented to the history section of the Royal College of Physicians and Surgeons of Canada and which later appeared in the first issue of the *Canadian Bulletin of Medical History*. He said that, despite what some people thought, Dr. Scarlett had been a basically humble person. If anything, in Gardner's opinion, he had been a little too humble.

He had a brilliant mind. His way of approaching everything [was] that you had to find the answer if the answer was to be found. But through this he was still a shy, quiet and humble person. He had a humility that was real. When he was set apart in a meeting, you might find everybody talking and having a drink and he would be sitting alone, quite happy, quite preoccupied with what he was thinking.

On the other hand, Scarlett was not the least humble about giving advice on certain things, as Gardner discovered when discussing music with him. He virtually demanded that his friend purchase certain Mozart records and insisted that he learn the particular way Mozart's music is catalogued — not by opus but by Köchel numbers.

Scarlett was much in demand as a speaker. He delivered the convocation address to the graduating class in medicine at the University of Alberta in 1940 and at the University of Toronto in 1950. In 1955 he gave the Shepherd Memorial Lecture at McGill University and, in 1957, the John Stewart Memorial Lecture at Dalhousie.

"He could wax eloquent and give wonderful speeches," Dr. Gardner says, but he also had "an ability to put people to sleep." His problem was a sense of occasion. He didn't always know when and when not to be erudite. When the fiftieth anniversary of the Calgary Medical Society rolled around, he was asked to give the after-dinner speech. It was, Gardner remembers, the perfect place for a fifteen-minute or so talk about the old days, "to highlight the evolution of practice and tell a few sprightly anecdotes about Stanley, Macnab or other notable practitioners. Instead, with the gathering of colleagues who had had a busy day and were full of good food, drinks and wine, and with a head table at which virtually everybody was over sixty-eight, he delivered an academic address which went on for about an hour and fifteen minutes." From what Dr. Gardner remembers of it, the speech would have been considered excellent by some bright and alert assemblage attending an important lecture. "He had that failing that he didn't seem to appreciate that that was the wrong occasion to offer that learned discussion." Gardner also remembers a fellow diner's more pithy expression: "Jesus! Is there any way of shutting him up?"

In the 1940s, Scarlett tried his hand at broadcasting. He was a member of a "round table" discussion group, but was more or less asked to leave after an argument over education. Scarlett was not exactly progressive. His suggestion that all city students be taught Latin was not well received. Around this same time his association with the University of Alberta began to pick up when he became a member of the Rhodes Scholarship Committee. In 1952 he became the first doctor to be chancellor of the University of Alberta, a position he cherished and held until 1958.

By the mid-1950s, and in his sixties, Scarlett found himself busier than ever. Aside from his duties as chancellor, he was president of the clinic and served on its administrative committee. By this time there were thirty doctors in the clinic and a large nursing and administrative staff. And he was still attending patients at the Colonel Belcher. He thoroughly enjoyed all his commitments. "I spent most of my life commuting but it was a lovely arrangement. Those were the days of the train. The golden days of life."[13] He took the night train to Edmonton, slept on the journey, attended his university board meetings the next day, caught the night train back to Calgary, arrived at 7 a.m., had breakfast and reached the hospital to do his rounds at the usual time. He liked to think of it as the "golden days" but physically it was too much for him to handle. He saw little of his family. An old friend, Dr. Morgan, finally told him he had to slow down, that it was time to drop many of his responsibilities and at least semi-retire.

He responded to Morgan in his own way. In 1956 he retired as president of the clinic and in 1958 cut back on his clinical responsibilities by 50 percent. Instead of slowing down, however, he was ready to devote more time to some of his avocations. Within a few years he would be known for his writing by physicians around the world.

Chapter Five

My mind, he said, rebels at stagnation.
— *The Sign of Four*

W HILE EARLE SCARLETT WAS SATISFIED with his position at the Associate Clinic, he was not completely happy with Calgary. He may have disliked the University of Iowa's materialism, but he appreciated the intellectual life of an academic institution. Before his Iowa sojourn, he had spent years in two even larger centres, Detroit and Toronto. Today Calgary is a sophisticated city with thriving drama groups, a professional symphony orchestra, opera and ballet companies, museums and a university. In 1930 the city's cultural life was mainly on an amateur level, and the nearest university was 200 miles to the north. Calgary had spirit, Scarlett admitted, and he did not want to leave the clinic, but he needed something else, some intellectual exercise which would remind him of the world outside.

He found part of his solution in supporting cultural activities, but what probably saved him was history. As McNeil said, Scarlett "loved the history of medicine, as he loved the history of anything."[1] As far as he was concerned, history was of common interest among what he called "the clerisy": those with the minds and the backgrounds to appreciate history and good literature. His medical colleagues, or so he believed, obviously fitted into this category. He saw the clinic as a suitable place to stimulate an interest in history, especially since one of the senior members, Stanley, had an equally strong interest in the subject.

The clinic already held a weekly clinical meeting, but in 1932 Scarlett proposed that the members hold an extra meeting once a month, an evening assembly at which papers devoted to medical history or some other paramedical subject would be given. It wasn't called a "historical meeting" at first, but the name gradually caught on. In time it became the more formal Historical Society.

Scarlett could be persuasive, but the meetings might have lasted only a short time had it not been for the support of one of the clinic's most important people, one with no medical degree – clinic librarian Frances Coulson. Mrs. Coulson was, Scarlett said, "an amazing woman," around whom the clinic revolved. Even Macnab used her as his adviser and mother confessor. She presided over one of the clinic's most civilized habits: afternoon tea. Each afternoon a bell would ring in the clinic and the doctors would gather in the library for twenty minutes to chat and drink tea poured by the librarian.

The historical meeting was held in the library and featured two papers delivered by clinic doctors who, as a rule, would have been less nervous doing a complex surgical procedure. Scarlett made sure there would be two papers by the simple expedient of grabbing a doctor in the hallway and saying something like: "Now listen Alex. I want an historical paper from you on such and such a date. Now you damn well get busy."[2] Oddly enough they did get busy, although, as Dr. McNeil noted and as Scarlett must have discovered, not all the papers were written single-handedly by the clinic doctors. Mrs. Coulson often turned them out herself.

Dr. McNeil was one of those caught in the hallway. He never forgot his delivery on "The Influence on History of the Illnesses of Henry VIII." According to McNeil, those given assignments normally worked very hard on them. But if, as in some cases, the author was seen pouring over what was his first paper since college days, Mrs. Coulson would step in. McNeil shared Scarlett's sentiments about her, describing her as "a very kind and considerate person, highly intelligent. A lady who loved books."

He remembers the librarian coming to him and asking, "How are you getting along with your paper, have you done anything?" To which he had answered, "No, but I'll get to it. I really must do it myself." McNeil had the best of intentions, but the fact is he kept putting it off. "Eventually Mrs. Coulson handed me a paper," he recalls. "It was beautifully done."[3]

Those attending the meetings would sit quietly and listen intently. In the early days it seemed Scarlett had indeed tapped a special vein of interest; soon doctors from outside the clinic were invited to listen and to give papers themselves. It was one way Scarlett hit on to overcome the professional barriers within the Calgary medical community.

As the meetings took hold and clinic doctors invited more colleagues from the general medical community, attendance began to rise. A few of the dragooned physicians hunted down photographs or portraits of their subjects, had them framed, and delivered their papers with the picture sitting beside them. This custom eventually resulted in the clinic collecting its own portrait gallery, and some of these pictures now are a part of the University of Calgary medical archives.

McNeil remembers medical pioneers from Grande Prairie to Lethbridge who turned up at the meetings to relate their experiences in the days when Alberta was still the frontier. The clinic paid their expenses and gave each an honorarium. To its credit the clinic bore the cost of operating the Historical Society.[4]

The society was one of the things Scarlett was most proud of. He presented not merely papers but people. Once each year he obtained from the registrar of the College of Physicians and Surgeons of Alberta the name of the oldest living physician in the province. He then called the doctor in question and said, "Now look here. This is your last chance to speak in public on this earth. I want you to come up."[5] The clinic would put the speaker up at the Palliser Hotel, and upwards of a hundred people would arrive for the evening's meeting. Scarlett's instructions to the ageing doctor were simple: "I want you to talk about your impressions of medicine, your experiences, your hopes for the future, and this is your chance to give advice to young men, because half the men in your audience tonight are young practitioners."[6]

Scarlett was not only delighted with the turnout at the meetings but also with the fact that, in many cases, it was indeed the speaker's "last word on this earth." Dr. Archer, president of the Canadian Medical Association, died two weeks after a large crowd heard his speech at the clinic. But best of all, within a few years Scarlett had ensured that at least some of these words would be preserved and spread throughout the Canadian and foreign medical community.

Scarlett had started the history meetings, but it was Stanley who came up with a suggestion which made the clinic unique in North America. One day in the mid-1930s, Stanley grabbed Scarlett, who had been standing in the rotunda by the clinic's front doors, and said, "Look here, Scarlett. I've just been doing a little thinking. We've got a good historical society here and I understand you're a writer. Why don't you start a magazine."

Scarlett protested that he had little experience as an editor, but Stanley wouldn't be denied. The older doctor said he would help gather material as long as Scarlett did the editing. More to the point, he would also make sure the clinic doctors coughed up the money needed for the project. "My colleagues," Dr. Scarlett later noted, "hadn't the remotest idea what the hell it was all about." They did, however, go along with the idea.[7] And thus was born the *Calgary Associate Clinic Historical Bulletin*.

In the mid-1950s Richard M. Hewett was searching for good examples of medical writing. He chose works by one Earle Scarlett from the obscure city of Calgary. "Long have I been a delighted reader of Dr. Scarlett," he wrote, and backed this up by reprinting an article entitled "Words and the Faculty" from the November 1953 edition of the *Bulletin*.[8]

In his essay "Dr. Earle Scarlett: Melding Tradition and Beauty in Historical Writing," Charles Roland, in 1980, noted that "with the possible exception of *le Docteur*, published in Montreal between 1922 and 1926, the *Bulletin* remains Canada's sole venture into medical-historical journalism. It is certainly our only anglophone journal of its type."[9] With the appearance of the *Canadian Bulletin of Medical History/Bulletin canadien d'histoire de la médicine*, this is no longer the case.

Scarlett's medical journal was not only unique, it was also free. For its twenty-two years the cost was borne by members of the clinic. The first issue, which appeared in March 1936, contained an editorial by Scarlett and a speech Stanley had given at one of the historical meetings. One thousand copies were printed and distributed to medical colleagues of the clinic's doctors. The journal's reputation grew with the list of contributors, and Scarlett found himself receiving papers from as far afield as New Zealand. What delighted him the most, however, was that his journal gave him a place to print the speeches of the medical elder statesmen he brought in to speak to the Historical Society. The one who did not get a chance to speak was Scarlett himself. By the time he qualified, the *Bulletin* was long defunct.

As for Stanley, he fulfilled his promise by becoming the journal's major writer until his death in 1954. He wrote essays, but more than that, he contributed information he had come by through his own personal connections. In 1948, for example, Stanley wrote about Dr. Abraham Groves, a surgeon who in 1883 performed his first appendectomy and in 1875 may have been the first in Canada to remove the uterus through the vagina, thus without the added operative risk in making a large incision into the abdominal wall. Groves may have also been one of the first to operate wearing rubber gloves. The gloves in question were those he normally used to drive his carriage.

The connection between writer and subject was Groves's son Billy, who was an intimate of Stanley's at the University of Toronto Medical School in 1901.[10] Often the connection was much closer. Stanley, after all, practised for half a century in Alberta and had worked with many of the men he wrote about. In himself, he was a historical resource whom Scarlett treasured.

In Stanley's case the *Bulletin* was the recorder of a pioneer doctor's final words. The last reminiscence, on Dr. J.F. ("Windy") Ross, one of his professors at Toronto, was completed only a day or two before Stanley suffered a heart attack which was to prove fatal. As Scarlett noted, part of his colleague's legacy was his written record.

> He and his contemporaries whom he celebrated for years in this bulletin, in the section called "medical pioneering in Alberta," created the traditions of the West which are now our heritage . . .
>
> In the last decade of his life he relived the years of his youth. He recreated the early days of the West, he wrote about them, he savoured them in the tranquility of the harbour of age.[11]

Stanley divided most of his work between two columns, the first on pioneers and the second on "Unforgettable Moments in Practice." In Scarlett's mind, Stanley "poured everything he knew into those two columns. The net result is that those columns give an authoritative, life-like, vital account of early medical practice in Canada, which you get from no other source."[12] He died in harness, from a coronary which occurred while he was talking to a patient. Scarlett's office was across the hall, but the older doctor was already unconscious when he reached him.

The 1,000 copies printed of the first issue of the *Bulletin* was a large run for a tiny journal published in a city stuck out in the prairie. The invaluable Mrs. Coulson found herself saddled with the job of making up the mailing list. Scarlett already had a large number of epistolary friends as well as colleagues in Iowa and Toronto, so he could contribute a good many names. He helped keep costs down by stuffing journals into envelopes himself, but, again, the clinic bore the brunt. Although the Depression had arrived, there were no complaints from clinic doctors about the cost. They simply figured Scarlett and Stanley were slightly crazy and they let them get on with it.

Volume 1, number 1, began with a note from the editor:

This number is the introduction to a series of articles which the Calgary Associate Clinic will issue quarterly to the members of the profession in Calgary and adjoining districts. Each succeeding number will comprise a brief résumé of the proceedings of "Historical Nights," held monthly under the auspices of the medical staff of the Clinic, and will endeavor to review in turn the lives and works of that galaxy of Pioneers in Medicine, who, by their foresight, industry, determination and self-sacrifice, have established during the centuries the eternal principles of the Science of Medicine, and bequeathed to this modern generation the great historical doctrines on which those principles are based.[13]

The first paper, "The Odyssey of Medicine," was subtitled "The Value of Knowledge of Medical History in Practice." It was written by Scarlett. History's value, he stated, was manifold. For the layperson, a knowledge of medicine and its growth is necessary for the understanding of how modern life is carried on. For the medical person, such knowledge ensures that the art of medicine is not lost in the welter of scientific facts, "a perspective from which Man may be seen as more than a walking test-tube or a co-ordinated machine."[14]

Typically, he backs up the heart of his argument by quoting Osler.

But the particular value of a knowledge of medical history involves other aspects of medical practice. A physician in his daily round needs two things beyond the cardinal necessity of a first class scientific and clinic training. He requires a general cultural background if he is to retain a broad outlook on life, and an inexhaustible fund of idealism if he is to give his patients the sympathy and courage they demand of the healing art. A knowledge of the great men of medicine and the spirit that animated them will give the harassed physician something of both. The leaven of their spirit will avert the fate that so often overtakes the practising physician – the fate that Osler describes as "the only too common nemesis to which the Psalmist refers: 'He gave them their heart's desire, but sent leanness withal into their souls.'"[15]

Scarlett wrote this as a physician who had had first-class scientific and clinical training and who, at that time, was ahead of most of his colleagues in Calgary in terms of experience and skill. His words reveal the inner man, a man who wanted to be a philosopher of medicine as much as a practitioner. He continued to publish the occasional paper on clinical subjects, but most of his writing from this time on was an attempt to shore up his colleagues' "general cultural background." His preference was not to preach at them, although he certainly did some of that – he was a preacher's son, after all – but to allow physicians of the past to send the message.

The remainder of the first issue was given over to Stanley. It included the first paper ever to have been delivered to the monthly historical meetings. "Peaks in Medical History" (a borrowed title) ranged through the ages from the pre-Hippocratic period to Canada's recent past.

The second issue contained a three-page essay by Scarlett, a paper on Leonardo da Vinci by W.A. Lincoln, M.D., and one on Oliver Wendell Holmes by H.W. Price, M.D. There was obviously a problem with the third issue, which contained little more than a six-page essay by Scarlett on the physician as philosopher. It was notable, however, for it carried the *Bulletin's* first illustration, a facsimile of a letter written in pencil by Florence Nightingale. The only other entry was a one-page appreciation, by Stanley, of Dr. Frank Hamilton Mewburn, a pioneer Alberta doctor and professor of surgery at the University of Alberta.

Number 4 got back on track with an essay on Thomas Sydenham by A.E. Aikenhead, M.D., and a short sketch of a still-living medical pioneer, E.A. Braithwaite, again by Stanley.

The clinic's other senior partner joined the action in volume 2, number 1. Macnab contributed an eleven-page essay on the founder of orthopaedic surgery, Hugh Owen Thomas. Macnab knew his subject well and had a deep appreciation of the physician and surgeon whose works he had so assiduously studied. He not only discussed Thomas's work but, like a good biographer, examined his character and life, from what he ate for breakfast to how he died from pneumonia after a house call.

In 1937 the *Bulletin* began to take its final form. There would be at least one long paper, a sketch of a pioneer doctor, usually by Stanley, and a column by Scarlett that was a cross between an editorial and an anthology of quotations.

In volume 2, number 4, Scarlett put a name to his column, one which not only summed up his main offerings to that time but also characterized his future published work. He called it "A Medical Miscellany," with the subtitle "From the Log-Book Book of a Medical Reader." In the column he quoted Sir Clifford Allbutt's *On Professional Education*, Dr. Samuel Johnson, Lord Horder (the king's physician), physician and writer A.J. Cronin, George Bernard Shaw's *The Doctor's Dilemma*, the nineteenth-century divine Sidney

Smith, Poet Laureate Dr. Robert Bridges and humorist Ogden Nash. Some quotes were long, others merely aphorisms, but the column established what would become a familiar format. In volume 3, number 1, for May 1938, he · changed the subtitle slightly to what would become almost a trademark: "From the Commonplace Book of a Medical Reader." Almost thirty years later, when his byline appeared in *Archives of Internal Medicine*, the title for his column would be "Doctor Out of Zebulun" and the subtitle, "Gleanings from the Commonplace Book of a Medical Reader."

The subtitles were important – "The Medical Jackdaw," the subtitle of his column in *Group Practice*, was a reflection of the "gleaning" subtitles because it illustrated not only Scarlett's way of writing but also a personal habit. Commonplace books were beloved by the Victorians, who would paste into these scrapbooks clippings from newspapers and journals, copies of favourite poems and any other trivia that took their fancy. To Scarlett, a commonplace book, which sometimes took the form of actual scrapbooks and sometimes of file folders, made perfect sense. Anyone entering his study could not help but notice, on a shelf on the back wall, piles of thick scrapbooks.

They struck the eye of William B. Bean, M.D., on the one occasion he met his favourite correspondent. "Early in life," he wrote, "Scarlett began keeping his teeming thoughts in commonplace books which have propagated like mushrooms. They have spread all over his library where, amid obvious chaos, he can put his finger on just the quotation, idea, or comment he wants. These provide the gems for the essay mosaics he has developed with such joyful artistry."[16]

Scarlett himself seemed less sure of the efficiency of his filing system. "When I was a boy at school," he wrote, "I began to note down or clip lines and phrases which caught my fancy. My motto was Captain Cuttle's, 'When found, make note of.' Gradually my habit became virtually a vice. Files and folders multiplied as the years went by until it became impossible to find some item which I wished to refer to."[17]

That complaint, written in 1944, was a bit modest. Even in the 1970s he would often stop the flow of conversation for a moment, put down his pipe, and walk over to some particular "commonplace" file or book. "This will interest you," he would say, pointing out some news item or piece of poetry which was germane to the conversation.

The notes from his commonplace books reveal the importance he attached to humour for putting one's point across. In the August 1938 *Bulletin*, he cited a story from Sir James Crichton-Browne's *The Doctor Remembers* about a medical student named Jones who was asked in an oral exam for the definition of an acute disease. "'There is no definition of an acute disease,' replied Mr. Jones. 'Sir,' rejoined the examiner, Professor Laycock, 'You have heard me say in my lectures that an acute disease is one that runs its course in fourteen days.' 'Yes,' retorted the imperturbable Mr. Jones, 'I have heard you say it, but there is no definition. An omnibus runs from Edinburgh to Leith in twenty minutes, but that is not a definition of an omnibus.'"

As time went on, much to Scarlett's gratification, articles from outsiders began to appear in the journal. Dr. Heber C. Jamieson, of Edmonton, wrote on Marion E. Moodie, the pioneer Alberta nurse; soon papers came from even further afield. Nor were all writers medical men. H.S. Patterson, K.C., contributed an essay on Sir James Hector, M.D., in the November 1941 edition; T.A. Reed, then financial secretary to the University of Toronto Athletic Association, outlined the first hundred years of the Toronto General Hospital.

Some of the essays would never make it into a medical or quasi-medical journal today. In the August 1944 edition, Dr. Walter S. Johns, a friend of Scarlett's who became president of the University of Alberta and who contributed the Scarlett entry in *The Canadian Encyclopedia*, wrote a delightful piece in praise of the "mystic Maiden Nicotine." Admittedly it did warn of tobacco's ill effects, but Johns impishly cited the case of the Rotterdam pipe smoker who died "prematurely" at eighty-one after smoking an estimated four tons of tobacco.

Scarlett, the inveterate pipe smoker, was another unabashed worshipper of the goddess nicotine but, like Johns, was not blind to her faults. In the February 1946 edition he noted that he had always wondered why the advertisements for Sweet Caporal cigarettes seemed to have the approval of the *Lancet*. He quoted from the advertisement: "The purest form in which tobacco can be smoked." But he repeats an excerpt from a *Lancet* editorial on the ads from a 1905 edition: "Bah, humbug." "So there you have it," Scarlett comments. "After more than forty years this particular firm goes on using a slogan that was coined in dishonesty and disowned at birth."[18]

One memorable essay, which may have seemed out of place to those who did not know Scarlett, was about an imaginary doctor, John H. Watson, M.D. In his introduction to Kathleen Morrison's paper on Sherlock Holmes's somewhat slow-witted friend, Scarlett admitted in passing that Watson was a "physician of fiction." He then, as a good Sherlockian, proceeded to write of him as though he were real. "To him we owe the record of the adventures of Sherlock Holmes, and in this role of biographer he must be ranked with the inimitable Boswell. Indeed it might be argued that Dr. John Watson is a more vital figure than many of the physicians who have been written about in these pages, and who more often than not seem like stone knights on old-world cathedral tombs."[19] Scarlett, naturally enough, allowed his Baker Street Irregulars' interest to creep in again and, in the November 1943 edition, wrote an essay on Dr. Joseph Bell, the man on whom Sir Arthur Conan Doyle based his detective.

On occasion Scarlett would venture out from his column and contribute a larger essay, generally on names from the past such as Lister or Jenner. By 1940 the *Bulletin* and Scarlett were well enough known that he was invited to deliver a paper to the seventh annual meeting of the Canadian Medical Association's section on medical history in Montreal. The subject he choose was Dr. Oliver St. John Grogarty, and the *Bulletin* reprinted the paper in two sections, in the November 1940 and February 1941 editions.

In the November 1944 edition he abandoned any pretence at history and penned an editorial on "The Physician and the Public," which reads as though it might have been written in the 1970s or 1980s. Respect for the doctor is fast disappearing, he wrote, and is being replaced by "impatient criticism and even veiled hostility." Part of this he blamed on the ignorance of the public, who "still think of diseases as primitive savages think of them – as so many evil entities each of which has its separate cure."[20] The rest of the blame he spread around.

The public at large has come to look on science as the great Alladin and many feel that the benefits which might be expected to flow from medical advances are not being given to society in anything like full measure. In part this is due to an antiquated scheme of medical organization, to the ignorance and folly of the public, and certainly in a measure to a narrow and reactionary attitude on the part of individual physicians. For no longer may medicine pride itself on being one of the "liberal" professions. The insistence on intensive specialized training has resulted in making the physician a better craftsman, probably, but an infinitely less educated man and a poorer citizen (using the term in the best sense of the word). Nor does it seem that much improvement is possible as long as medicine remains a commercially competitive trade. The physician under the compulsion of modern forces has joined in the economic scramble, making it more and more difficult for the guiding humanitarian principles and the great tradition of medicine to maintain their influence. That is the dilemma of medicine at the moment.

And the solution? There is no precise answer to this question. Certainly under the impact of war and social changes a more effective organization is coming. The medical profession itself may in some degree guide such a new order. But its first duty is in another sphere. The quality of medicine depends on something other than organization. It depends on its standards, its sense of values, its spiritual outlook, its broad point of view based on education. If these are of the best it is much less important what social form medicine will take, for the medical profession will then be assured of the abiding faith of the public.[21]

The anxieties he outlined in those two paragraphs, and in the rest of the editorial, he carried to the grave. Sadly, he felt that the solution had eluded his profession and that the dilemma's horns had only grown sharper over the years.

The *Bulletin* was normally printed on rough but good paper stock in blue ink. In 1945, its tenth anniversary, the journal was printed in black ink on coated stock. The 106-page issue contained a fascinating collection of papers from both Canadian and American writers. They included Dr. Henry E. Sigerist, professor and director of the Institute of the History of Medicine at Johns Hopkins; Dr. W.B. Howell, former chief anaesthetist at the Royal Victoria Hospital in Montreal; Dr. Howard Dittrick, of Cleveland, Ohio; and Dr. H.E. MacDermot, editor of the *Canadian Medical Association Journal*, to name only a few. And, of course, Stanley and Scarlett. The subjects ranged from a humorous history of hospitals to medical education. Scarlett's contribution was on a subject which was one of his passions, John Keats. Scarlett had already made a pilgrimage to Keats's London home, now a museum, and was able to examine the poet's anatomy notebook. He delighted in Keats the poet, but couldn't help admitting that the notebook showed that Keats couldn't spell and, judging from the marginalia, did not always pay attention to the lecturers.

From 1947 until 1952 the *Bulletin* presented short histories of Canadian medical schools, beginning with Dalhousie's and ending with the University of Saskatchewan's. Perhaps Scarlett's favourite series, besides Stanley's "Medical Pioneering in Alberta," was "Early Surgeons of the North West Mounted Police" by Dr. J.B. Ritchie of Regina.

Scarlett opened the February 1957 edition with an editor's comment which said, in part, "We had almost despaired of finding anyone to carry on

Dr. Stanley's work until by a happy accident last summer we came upon a solitary figure in the Stanley Library of the clinic busily writing at one of the study tables."[22] The figure, Dr. Ritchie, was an old medical friend of Scarlett's and he had been discovered researching the Mounties' medical history. He soon agreed to turn out a series of articles. One of the tragedies of the journal's demise one year later was that this occurred before Ritchie's series could be completed.

Of the non-medical outsiders who contributed many of the *Bulletin* essays, one name should be noted – George F.G. Stanley. Stanley, whose *The Birth of Western Canada* is one of the seminal works of Canadian history, was a reader of the *Bulletin* and, oddly enough, a man whose career was guided, like Scarlett's, by the advice of Prime Minister Bennett. The Calgary-born Rhodes scholar has written a great deal but none of his pieces is more delightful than "A Strange Remedy: Being an Account of a Gentleman of Lyons of his Experiments in Curing the Taenia or Tapeworm in 1750." The cure involved swallowing lead hooks attached to thread and fishing the worm out, an operation which worked, but which has not become popular among modern physicians.

Volume 22, number 4, the February 1958 issue, began with the heading, "*VALE.*" Dr. Scarlett had managed to continue as editor while serving both as head of the Associate Clinic and chancellor of the University of Alberta, but in 1956 he decided it was time to retire from active duty at the clinic and in 1958 he gave up his job as its head. He and Stanley, with Macnab's support, had begun the journal and he was the only founder left. Mrs. Coulson had retired, to be replaced in the clinic library first by Mrs. Aileen Fish and then by Mrs. Bernice Donaldson. Worse, only a few people at the clinic seemed to care about medical history. According to Scarlett, the clinic doctors were getting tired of the *Bulletin.*

Finally, Dr. Harry Morgan, Scarlett's successor at the clinic, came to him and said, "Earle, you're going to kill yourself. You're doing too much." Not only had Scarlett shouldered the responsibility himself for too long but, Morgan added, "the fellows are beginning to think that they're sort of subsidizing medical history all over the place."[23] Getting rid of the *Bulletin,* Morgan said, would take a large load off Scarlett's back.

Scarlett, who was thinking of celebrating his term as chancellor by taking a year-long holiday in Greece, finally agreed.

"With this issue," he wrote, "the *Bulletin* comes to an end after 22 years of continuous publication. No journal of its kind has attained such an age in this century. In some eighty-seven issues with the enthusiasm of the amateur and innocent of the more severe disciplines of history, we have dealt with all phases of medical history and have ranged widely into para-medical channels of past and contemporary thought and opinion."[24]

The final volume began with a paper by Dr. H.E. MacDermot of Montreal, a reader of the *Bulletin* since its founding and a frequent contributor. He discussed the careers of two medical women, Miss Nora Livingstone, founder of the Montreal General Hospital School of Nurses, and Dr. Maude Abbott, who successfully fought for her admission to the McGill University Medical School in 1890 and then became a cardiologist at the Montreal General. This was followed by the recollections of Dr. David M. Baillie, medical superintendent of the Gorge Road Hospital in Victoria, B.C., and a paper by Dr. E.H. Bensley, director of the Department of Metabolism and Toxicology at the Montreal General, titled "Skulduggery in the Dead House." This dealt with the difficulty medical students had in obtaining suitable subjects for anatomical dissection. The final paper was Dr. Ritchie's fifth chapter in his history of the North-West Mounted Police surgeons.

And, of course, this last issue ended with a dip into the "Commonplace Book of a Medical Reader." For this last effort, Dr. Scarlett pulled notes out of a file folder marked "Finally," a title he notes was inspired by St. Paul, "who was a great writer of 'finallys' and whose supreme injunction in this manner is recorded in *The Epistle to the Philippians* (Chapter 4, verse 8)."[25] Other passages he quoted are from writers as diverse as Aristotle, Walter de la Mare and, of course, Hippocrates: "Words which echo down the centuries for all medical pilgrims . . . 'Life is short and the art long, the opportunity fleeting; experiment and judgement difficult.'"[26]

Though perhaps a relief of sorts, the loss of the journal, nonetheless, hurt Scarlett, and not merely because something he had conceived was dying. Looking back on his experiences years later, he may have thought he had failed. The *Bulletin* was to have been a bulwark against materialism, but he felt it had not stood the test of time, even in his own clinic. "As far as medical history is concerned," he told Roland, "you might as well talk about having Sanskrit parties."[27]

His despair was not justified. He had gathered up the stories of early medicine, including many stories which might otherwise have been lost, and had turned them into a permanent record for the benefit of those who would share his love for medical history. He had left a legacy that would inspire others with similar convictions. Dr. J.S. Gardner, in his appreciation of Dr.

Scarlett in the *Canadian Bulletin of Medical History*, says, in effect, that Scarlett did not fail, as he would himself have agreed had he lived to see the first issue of this new Canadian journal, one in direct line of succession to his own. It was fitting that the first issue of the *Canadian Bulletin* began with a photograph of the title page of Scarlett's own journal.

In Charles Roland's words, everything about Scarlett's journal was "unlikely." It appeared in the middle of the Great Depression in a small city in the middle of nowhere and was funded by a group of doctors who weren't even sure what they were paying for. Perhaps the most unlikely fact of all was that it lasted more than three decades.

Scarlett might have chosen for his journal's epitaph that one "finally" quote from St. Paul:

> Finally, brethren, whatsoever things are true, whatsoever things are honest, whatsoever things are just, whatsoever things are pure, whatsoever things are lovely, whatsoever things are of good report; if there be any virtue, if there be any praise, think on these things.
>
> — *Phil. IV 8*

Chapter Six

It was a large and bright dwelling,
rather a villa than a cottage . . .
— "The Adventure of the Devil's Foot"

EARLE SCARLETT LOST NO TIME in putting down roots in Calgary. He and his wife arrived with one son and soon there was a daughter, Elizabeth Ann. She was born in 1931 and a second daughter, Mary Katherine, followed in 1935. But while his children may have provided for posterity, they did not necessarily provide the one thing the young internist needed – stability.

Scarlett was proud of being a child of the manse, but the manse in his case had not been a solid, ivy-covered residence beside a picturesque church. His had been a movable manse, a peripatetic household which marched back and forth across the prairies. Perhaps because of this, he needed and sought permanence, a place of stability, an emotional centre to which he could return and from which he could draw strength.

He found the stability he was looking for in his wife. Her down-to-earth Scottish temperament seemed to provide a perfect counterpoint to his romantic, Celtic moods. She was a practical woman, but she shared his love for music and literature and had been an award-winning scholar in university. From an intellectual point of view, she was a perfect match for him.

As a doctor's wife, she knew his practice came first. If she was saddened by this, her family had no inkling of it; it was what she expected. At least he talked to her about his work and allowed her access to the world which demanded most of his attention. Between his practice, his reading and his writing he had little time to spare, but she was still the touchstone of his life, his accountant, his chief critic and his patient listener. Her work in the home gave him the freedom to spend time on his practice, his books and his pen. As an Edwardian – if not a Victorian – gentleman, he took her activities for granted, but the tributes he wrote to doctors' wives in various journals were aimed at her.

Their children look back on the strength of this relationship as something they took for granted but which now they treasure. Robert calls attention to the dedication written in one copy of his father's unpublished literary anthology "The Ram's Horn." It reads: "To Jean: The beginnings of an indissoluble relationship, altogether remarkable in my experience."

As a child, Robert says, he thought all marriages were this way."If they ever had a serious quarrel, they certainly kept it well hidden from the family. The only disagreements I can remember were relatively minor: my mother would get quite upset at my father's ability and inborn inclination to embellish a story, and exclaim times without number, 'Oh, Earle, it didn't happen that way!', in tones of increasing exasperation as the story went on."

If nothing else, their mother's comments taught them to be sceptical about the literal truth of their father's stories – not that they enjoyed them any the less.

As Kay points out, that was only one of the dedications for "The Ram's Horn." The original dedication, written in 1944 when Scarlett began collecting material for the anthology in a formal way, ran: "To Jean: The Owl makes amends to the Lark."

Earle was a nighthawk, Jean was not. In the evening, when he came home from a house call or a hospital visit, he would drop into their bedroom and check to see if she was awake. If she was, they would talk for a while, and then he would disappear into his study and stay there often long past midnight, writing, reading, clipping items for his commonplace books or his anthology.

Years after beginning "The Ram's Horn," perhaps even after Jean's death, he added these words to the dedication:

As I look back I can see that this book was inevitable from the beginning. You, of course, knew all the time. In the university year-book at the time of my graduation in Medicine there stands under my name a line which you chose: "He would hunt half a day for a forgotten dream." Blessings on you, my Lady! And now these are the records of many a half-day's hunt.

Betty remembers her parents' relationship in terms of her mother being the rock on which her father's life was built.

> She ran the household, looked after mundane affairs, saved Father from having to deal with trivialities, listened to his stories, commented on his writings which he would read aloud to her. Their relationship was something very special, which made it all the more difficult when she died in 1975. In fact, he put together a book of writings – a small anthology called "A Breviary of Love." She was unassuming, gentle, wise – supported father in the limelight, especially in his days as chancellor of the university – yet also had a sparkle and wit of her own, and an ability which showed up in her positions on the Calgary School Board and library board.

Their father didn't talk much about his own antecedents, but Jean wrote her young children a short account of her childhood in Cobourg, of her mother's death when she was six, and of the aunts and uncles who raised her. At the same time, they say, she was self-effacing, seemingly content to be overshadowed by her husband. She was the chatelaine of his castle, a large frame house at 409 Roxboro Road. It was this house which became the physical centre he had lacked as a child. As physical infirmity shrunk his world, it became even more important, something firm he could cling to after Jean's death. It was, in his words, the "Kingdom of 409."

One woman who knew her well, Mary Cairns, noted that Jean Scarlett did not talk about being a doctor's wife; she took her status as a housewife for granted. "Dr. Scarlett had nothing to do with the running of the house," Mrs. Cairns says. "That was the way it was: keeping house was the wife's job. She was perfectly happy doing her job as a housewife and taking part in community things. The kitchen was the most appalling kitchen I have ever seen. The size of it and inconvenient! The only table space at all was a tiny little counter to the left of the sink. Why she put up with it I'll never know." But the rest of the house was lovely.

Jean played the part of the traditional professional man's wife with competence and spirit, putting on tea parties for the wives of new doctors, playing host to a woman's "sewing circle" and a separate "literary circle," and managing dinner party after dinner party. Sometimes her husband's behaviour annoyed her. The roast or turkey would be brought in, and Earle, intent on the conversation, would ignore it until the meat grew cold and Jean would be forced to speak up and order him to get on with his duties.

But she was more than a good hostess. She was active on some of the city's most important boards. She served on the board of the public library from 1949 to 1959 and was chairman for the last three years. She also won a seat on the school board. Those who were her colleagues praise her effectiveness for any cause she chose to champion.

"I remember my mother as an extraordinarily independent person," Robert says. "She did everything that women were permitted to do at the time in the way of political activities; it wasn't much, but she made the most of it."

Sitting on the library board or the school board may sound conventional, but as a person with strong opinions, she was quite willing to be controversial when necessary, even if it meant taking on the medical establishment or the Roman Catholic Church.

In the 1930s she became interested in birth control, a cause which was still controversial and which was considered by many to be an improper interest for a lady. On 7 March 1939, at a meeting held, fittingly enough, in the Young Women's Christian Association (YWCA) building, Jean was elected president of the Calgary Birth Control Association.

Birth control was considered by many to be unnatural at best and positively anti-Christian at worst. Even the legality of disseminating birth control information was in doubt. Mrs. Scarlett considered it a public health necessity and wasn't just satisfied with speaking up for birth control: she led her troops to city hall to fight for the hiring of a social worker who would work as the city's "birth control officer." Her specific proposal was that the YWCA be given a grant that would allow it to hire such a person.

The Roman Catholic bishop of Calgary was aghast, and even Calgary's medical health officer, Dr. W.H. Hill, announced he was against any such idea. If birth control information was to be given out, he said to city council, it should be a matter between a woman and her private physician.

Jean Scarlett, in putting her case to the public, pointed out to a reporter for the *Calgary Herald* that Dr. Hill's own 1937 report on city health had stated that six women had died after an abortion. "We have no means of knowing exactly how many of the abortions were an attempt to circumvent the result of inadequate preventative measures, but we believe that this factor played a significant part and for that reason the figures are significant," she said.

On 3 April, after listening to both Jean Scarlett and Dr. Hill, not to mention the receipt of numerous letters for and against the idea, the city council voted against giving such a grant to the YWCA. Instead, it voted to hire its own birth control officer, "a private nurse or such other qualified person being a female."

Alderman James Mahaffy, showing sentiments ahead of his time, noted that while it was fine for Dr. Hill to say this should be a matter between a woman and her doctor, "there were many who could not ask doctors for it."

An editorial in the 4 April 1939 edition of the *Herald* noted, "With this new official on the job Calgary will be the first Western city of the Dominion officially to recognize municipal responsibility in the matter of birth control. The progress of the movement will undoubtedly be watched with much interest by the citizens generally."

There was, as could be expected, an outcry over the council's decision, but Jean Scarlett's stand had worked a change in the medical history of Calgary. (In the 1980s the outcry has been just as loud and even more angry when city health officials admitted they were giving out birth control information to teenagers.)

Both Earle and Jean Scarlett believed wholeheartedly in giving time to good or, at least, civic causes. These tended to follow their interests in music and literature. Scarlett became a board member of the Calgary Symphony Orchestra and kept up this connection for three decades. He, like his wife, was a member of the library board, a function which served him well, since, after his heart failure had precluded visits to the library, librarians made sure that any new book which might tickle the Scarlett fancy found its way to 409 before it hit the library shelves.

While the children loved and respected their father, they adored their mother. She was always there when they needed her as someone to confide in, to weep or laugh with. "We all know how much 'Grandee' was admired and respected," Betty notes. "We were known as 'Dr. Scarlett's children.' This had its good points, of course, but at the time we would have liked to be just ourselves. Maybe we thought too much would be expected of us.

"We always felt very proud and very secure at having such a man for a father, but there were drawbacks. So often he was not home. Even Sunday mornings and Christmas Day he was doing rounds at the hospital or house calls. He always seemed to be 'on call,' out in the evenings, again for rounds, or meetings, or, if not out then buried in some sort of reading. As a father, in some respects he was 'never there,' always so busy. At home he was immersed in his own reading and writing, or talking about his own interests. They were fascinating and we absorbed an immense amount of knowledge, but there were times when we too would have liked to be heard.

"He never suggested we go into medicine," Betty says. "But then he never suggested we do anything specific, we were to find our own way. We all 'did our own thing' at 409. I do wish, however," she adds, wistfully, "that father had had a little more time to do things with us."

Scarlett knew Betty enjoyed the outdoors and loved horses, and he made sure she had a succession of them. Robert had a chemistry set, with which he almost burned down the house. When he developed an interest in electronics and received his ham licence, his father allowed him to buy what he needed, including parts for a home-built transmitter. Kay's interest was artistic, and because her father loved Canadian art and knew many of its artists, he supported her as much as he could.

But at the dinner table, while questions were asked about their school-work and activities, it was the father who dominated the conversation. What he wanted for his family was stability. Family life was built around certain rules. Meals, even when their father could not be there, were formal by today's standards, and the children were expected to be at table on time. The one relaxed meal, Sunday-evening supper, was served in the living room from the tea cart. This, "the family hour," was the time the children remember best. Late Sunday afternoons were reserved for the great radio comedians – Edger Bergen and Charlie McCarthy, Jack Benny and Fred Allan.

The children, whether they wanted to or not, were required to attend this family hour. They would listen to music, sometimes recorded but more often homemade. Jean played the piano while everyone else sang carols or hymns. Then came the reading aloud. Jean would read passages from the Bible, and Earle would follow with children's classics, such as *Wind in the Willows* or *Alice in Wonderland*, and Dickens. "We got through a great quantity of Dickens over the years," Robert remembers. The children picked up their parents' appreciation for literature and for the literary worth of the Bible. Jean might also include something from her favourite author, Jane Austin.

"I would not have admitted it at the time," Robert says, "but if it had been interrupted, I would have missed it greatly. A fixed anchor of stability would suddenly have become unloosed."

Either because he felt a doctor's world was private or because he wanted to protect his family from the tensions which medicine creates, Earle Scarlett liked to believe that he did not bring his work home with him, that the imaginary walls around his castle at "409" kept the pains of his profession at bay. This was, of course, a delusion, but from their earliest days all three children recognized that their father was engaged in some sort of important and mysterious work.

The most the children knew about their father's practice came from the odd comment about so-and-so visiting the office or from one-sided snatches of a telephone conversation. "The phone would ring," Betty remembers, "and we would hear 'the doctor' speaking, hear him giving prescriptions to the pharmacist in an incomprehensible language that was always fascinating. But," she adds, "we never really knew how father felt about being a physician in a day-to-day sense. In anything personal he was a very, very private person."

That Dr. Scarlett could talk to his wife about his work was one of the things that made her essential, Kay believes. "Over dinner at noontime he would unburden his heart about some medical case or anguish over someone dying. (We were solemnly sworn to say nothing of what we might hear.) When very little I was impressed by the gravity of such things and for years felt anxiety at the ringing of the phone, for when I was small it was often a distressed patient and I had to be very careful to at least get the phone number correct."

Robert, perhaps because he was the eldest, remembers more, such as the numerous times his father would come down to breakfast and complain that "Mrs. X" had dragged him out of bed at 2 a.m. for some phantom complaint. When Scarlett thought the children might overhear, it was always "X" or "Y." Patients' names were not to be mentioned, although somehow, in a way Robert could never figure out, his mother always seemed to know who his father was talking about.

There was one exception to the taboo on mentioning names or describing medical matters. When Scarlett lost a patient, he would mention the person by name and give details of the case. Robert believes this might have been his father's way of coming to terms with the death of a patient or even death itself.

"At no time was this more striking than when my mother died. She was in a deep coma for several days, and he gave us a lucid and dispassionate account of her state, almost as if he were addressing a class of medical students."

In spite of his studied attempt at objectivity, he never became reconciled to the death of his wife. "In the end," Robert says, "that was one outcome he could never accept."

Long after he retired, when he was in poor health, he still thought of himself as a physician. He would tell family and friends that he could diagnose his own condition better than any of the "young sprouts" who were looking after him. He was quite ready to argue about the proper dosages of whatever drug he happened to be taking at the time. His pride in being able to do this helped keep him alive.

While the "family hours," which his children remember with so much affection, coloured by nostalgia, always included Bible reading and often hymn singing, Earle Scarlett was not interested in ensuring that his was a conventionally religious home. Jean, however, wanted her children to have some religious background and ensured they attended Sunday school and services at Wesley United Church. They took to attending church as a matter of course, just as they took it for granted that their father would not accompany them. His formal excuse, when he needed one, was that he had hospital rounds to do. The fact is that he had no intention of attending services and the only times his children ever saw him in church were at weddings and funerals.

Jean also taught them about money, whose management was not her husband's strong point. "My mother managed the household finances, and very prudently too," Robert says. "We were constantly exposed to such maxims as 'a penny saved is a penny earned' and 'it isn't a bargain if you don't need it.'"

Earle Scarlett did not go to church, but he had many friends among the city's clergy, especially among Roman Catholic priests. This would have horrified his ancestors, but Scarlett thought the priests had the best stories. He liked nothing better than to sit down with them and swap tales, tall or otherwise. While church-going did not interest him, faith did. He was haunted by what he perceived as people's lack of faith in humanity and in a higher power. In "The Ram's Horn," Scarlett states that "the modern existentialist maintains that the world and human destiny are a mass of incongruities, and, therefore, may be labelled as 'absurd.'" In disputing this, he quotes from Reinhold Neibuhr's *The Irony of American History;* Neibuhr, whom Scarlett calls a philosopher rather than a theologian, notes that the "final wisdom" of life is to achieve serenity and that, in the end, humanity is saved by faith, hope, love and forgiveness. Scarlett follows this quotation with the poem "There Is Nothing There but Faith," by Edwin Muir, and the comment "In the unenlightened level of common life there is the redeeming presence of faith and a glimpse of a hidden reality."

Those words, according to Robert Scarlett, sum up his father's religious convictions. "If being religious means having a belief that there is more to life than can be comprehended by our five senses, that there is something beyond our present existence towards which life is tending, and that a future life, no matter how incomprehensible, is less incredible than that there should be nothing at all, then I would say my father was deeply religious."

His was a religious humanism. It expressed itself in his tremendous faith in the spirit of man. Perhaps it was because this faith was so strong that he was distressed by anything which detracted from the dignity and worth of humanity.

Earle Scarlett's study was a place of delight. Visitors left alone for a few moments, generally while their host was digging out some book from his "stacks," could examine books, his pipe collection, and various pieces of memorabilia, some medical but many literary. One small, almost unnoticed item was a copy of Robert Louis Stevenson's "Requiem," a facsimile in Stevenson's own handwriting which Scarlett had bought in Banff in 1920 and had professionally framed in Calgary. The passage he underlined says a good deal about Scarlett:

> Here he lies where he longed to be;
> Home is the sailor, home from the sea,
> And the hunter home from the hill.

Scarlet was a man who enjoyed travelling, whether tramping about Greece with his wife or attending a medical convention in a distant city, but he needed his castle, his kingdom of 409. His family still talks of "409," although it no longer exists, rather than "home."

But there is another number which remains imprinted indelibly in the family's memory. This was "Number Six."

Scarlett had chosen Calgary partly because of its career possibilities but also because, as a CPR conductor, he had fallen in love with the mountains. Smitty Gardner, a prodigious hiker and skier who could have been a mountain guide as well as a surgeon, remembers that if he mentioned a pass or a trail, Scarlett was sure to have been there. He very early made sure his children shared his love for the mountains. His first vacation spot was a cabin by Kootenay Lake, near Nelson, British Columbia, and the second was a similarly rustic cabin on Ghost Lake, a reservoir created by Ghost Dam on the Bow River near the entrance to the mountains. There Betty fell in love with horses and Robert learned how to split kindling and fill the kerosene lamps.

They needed the lamps. Part of the family baggage was a box of books, and Scarlett burned kerosene late into the evening, reading by the same light he had used as a child.

84

In 1940, however, when Earle and Jean found themselves searching for a place to stay in Lake Louise, they happened on a brand-new set of tourist cabins, Temple View. One of them, Number Six, was completed, and the owners set up a bed for them. From that time, "Number Six" had almost the same cachet as "409." In his "breviary," Scarlett writes:

> This beautifully-designed cabin seemed to open its arms to us. It was one place where one could get complete rest, forget all the troubles and worries and be at peace. To stand in the doorway and look up at snow-capped Mount Temple was a never-failing inspiration. It was the setting for the peak period of our Family when we were all together. It witnessed the golden days of youth, of parental pride, of love and happiness.

The village of Lake Louise in the 1940s was still a primitive spot virtually unspoiled by tourism. The Trans-Canada Highway was well in the future. The highway west from Golden to Revelstoke, the notorious "Big Bend," was a frightening and chassis-pounding gravel road. A thousand feet above the railway station on the shores of the lake itself was the Château Lake Louise, an outpost of civilization built by the CPR. There wealthy tourists, who generally arrived by train, were treated royally. After a day's walking, they dressed formally for dinner, then swept down the grand staircase to the main lobby to listen to a concert pianist. Afterwards, late into the night, they danced to the music of Moxie Whitney's big band or some other professional orchestra imported from Toronto.

Down on the valley floor, in Number Six, dinner was cooked on a wood stove and the evening's entertainment consisted of conversation and reading. The lights were powered by an uncertain generator shut off at 11 p.m. Number Six became a symbol of the family leisure, and how they would all groan when they spied Calgary on the horizon on the drive home!

Temple View Camp, later renamed (much to Scarlett's disgust) Motel Lake Louise, had about a dozen buildings, and Number Six was slightly detached from the others. Earle and the owners became friends and it was understood that when the Scarletts wanted it – all of August and the occasional weekend – it was theirs. "It was a magical place, and perhaps one of the most enduring family traditions. The number came to stand for the entire country for miles around; we never went to Banff or Lake Louise, it was always simply Number Six."

It was a solid, two-bedroom cabin with a large fireplace dominating the front room. There they would even play cribbage, something Earle would never do back in Calgary. Without a phone or demands from patients, he would build a roaring fire or, wearing a thick flannel shirt, step outside for a quiet pipe in the cold mountain air. The sense of peace one can only find in such a place provided him with some of the most precious moments of his life.

Occasionally the five would drive east to vacation at an old family spot on Georgian Bay, but to the family a holiday was Number Six above all. "The mountains were knit into our family fabric," Katherine says. The children dubbed Jean the "mountain poet" because of her ability to make up rhymes as the family hiked along.

When Betty pictures her father in her mind the first image is "Grandee" sitting in his study or at his desk, wreathed in pipe smoke, deep in some medical journal or busy with his pen. The second is her father feeding the dogs, cutting up bits of lettuce or other vegetables to mix in with the dog food, or taking the dogs for a walk. Friends who lived close by remember that almost until he died he could be seen walking slowly through the tree-shadowed neighbourhood, his dogs beside him. The Scarlett dogs came in all shapes and sizes, from Paddy the Newfoundland, who pulled the neighbourhood children around in a wagon, to Topsy, a supposedly registered scotch terrier who lived to eat – anything from ice cream to an entire cooked chicken.

Scarlett wrote about his beloved animals in an unpublished anthology:

When I remember the long line of dogs that have been members of this household for the past 40 years, a host of memories comes crowding in, and the rarest kind of happiness is evoked – a happiness that is as pure and as unclouded as any this contrary world can offer ... Their ghosts throng our back garden. We have known the joy of being owned by dogs, the misery of reconciling the cruel fact of nature that a dog's life span is so much shorter than a human's (our greatest complaint, I think, about the way in which the President of the Immortals has arranged this universe), the incomparable fellowship that exists between man and dog.

Both Scarlett and his wife were busy people. He was able to juggle professional, social and intellectual activities because of a knack for organizing time, an ability to read quickly, and an excellent memory. When something came up, Earle and Jean could usually find time to squeeze it in.

Their shared love for music led them to meet with friends to sit silently and listen to records. These musical evenings became a tradition which lasted for more than thirty years. According to Judge Ted Tavender, a group of four couples began meeting informally back in the 1930s. Eventually one couple dropped out and the Scarletts were invited to join. Informality was the keynote until it was the Scarletts turn to play host. For the first time, a printed program was provided. From then on, 1 December 1940 was marked by Dr. Scarlett as the group's anniversary.

The host for each evening would prepare the written or typed program and distribute it to the audience. The music was listened to with respect, Bert Cairns remembers, and then all adjourned for light refreshment.

The difference between an evening at the Scarletts and an evening with the Tavenders, the Leslies, the Cairns or the Worralls was in the nature of the printed program. The others, Judge Tavender says, were content to list the evening's entertainment. Scarlett's program included information on the piece, the composer, and anything else of interest. These printed sheets grew even weightier for special occasions; the fifteenth and twenty-fifth anniversary programs, for example, included historical notes:

> The Group, not being an incorporated institution, has survived several changes in its title – at first "The Southern Alberta Mutually Exclusive Drama, Art, Literature, Music and Singing Off-Key in the Bath-tub Cooperative Society"; then "The Group of Eight"; then "The Music Group"; later "The Interlude Music Group"; shortly afterward "The Rejuvenated Music Group (1952)"; and finally "The Music Group" again.

The concerts ended, needless to say, with "God Save the Queen" – "God Save the King" in the earlier days.

"As a child," Robert Scarlett recalls, "I was fascinated by this convocation of grown-ups who would sit around and listen to records interminably." Gradually, however, the doctor's son grew up to have both a vocation, electronics, and an avocation, music. Robert, in the mid-1940s, contributed to the group by building what came to be known as "the machine," an early hi-fi set with dual turntables and an elaborate sound system. His father was proud of his son's creation and refused to part with it until the 1970s, when Robert provided a stereo which included FM radio. It sounded twice as good as the old set, in Robert's opinion, but he could never persuade his father of this.

The Scarletts passed on their love of music and literature to their children, who, in discussing their parents, bring up title after title of favourite books. One series they could not ignore was the adventures of the world's greatest consulting detective, Mr. Sherlock Holmes.

Scarlett's literary mind and his sense of humour were both touched when he read about the Baker Street Irregulars in the *Saturday Review of Literature* This group was founded by *Saturday Review* contributing editor Christopher Morley in 1934. It followed the lead of Mgr. Ronald Knox, who in 1919 wrote a paper titled "Studies in the Literature of Sherlock Holmes" as a parody of the heavy-handed German literary criticism then in style. Knox had treated Watson as the living biographer of an actual detective, and the Irregulars ran with this idea.

Morley introduced the Baker Street Irregulars to the pages of his journal early in 1934. Scarlett must have seen the articles, for Morley replied to Scarlett's "friendly and amusing letter" in May of the same year. For some reason, however, Scarlett wasn't moved to join the Irregulars, although he kept up his correspondence with Morley, who joined the list of those receiving the *Historical Bulletin*. Morley not only wrote to him on occasion about Scarlett's own journal but also, since Scarlett appreciated George Bernard Shaw, sent him a bit of doggerel he wrote for GBS's ninetieth birthday.

What obsequies for dear old Shaw
Who lived outside the canon law?
Let's give him, to be truly Shavian,
All rites, including Scandinavian.

In 1942, while rereading some of the stories, Scarlett decided he might use Holmes and Watson as a topic in one of his speeches. He asked a patient, Miss Kathleen Morrison, to research the subject for him. The first result was a speech Scarlett gave to the Edmonton Academy of Medicine. That body, expecting the usual clinical presentation, was "politely shocked" at a dissertation on an imaginary medic. The second result was the appearance of Miss Morrison's paper in Scarlett's historical journal. She sent a copy of this essay to the noted Sherlockian Vince Starrett, who passed it on to a member of "The Scandalous Bohemians" chapter of the Irregulars in Akron, Ohio. Both Scarlett and Morrison soon found themselves members of the Akron chapter, and Scarlett ended up a member of "The Baskervilles" of Chicago, his membership signed by "The Needle," Vince Starrett.

Morrison, a city librarian, found herself making minor international news. The Associated Press picked up a note about her membership and, on

29 January 1945, sent out a wire story calling her "the only active woman member of the many societies devoted to the lore of Sherlock Holmes."[1]

"It is all a piece of delicious foolery with that right mixture of fantasy and nonsense and love of literature that makes it irresistible," Scarlett wrote in "The Ram's Horn." "For its members can now revisit at will and share a pipe and good talk with others in a world where, in Vincent Starrett's phrase, 'It is always 1895.'"

Scarlett loved the "foolery" but, like Morley, he saw meaning beneath it. As he writes in his anthology:

> The members have adopted the convention that Holmes and Watson are real in the sense that deeper reality attaches itself to great symbols and great legends. For not only are Sherlock Holmes and John Watson in the realm of the living with Don Quixote and Sancho Panza, with Crusoe and Friday, and even with Dr. Johnson and Boswell, but they are central figures of the only authentic great legend that has been created in modern times.

His words echo those of Morley, who wrote that Holmes and Watson "have the reality that accrues only to the greatest symbols and cartoons."[2]

Shortly after World War II, Morley visited Calgary and Scarlett. He was rewarded with a ride to the mountains in a CPR locomotive. In a note thanking Scarlett, Morley also mentions their visit to "the 2nd-hand bookstore," and underlines the phrase. Scarlett, having read Morley's 1919 novel *The Haunted Bookshop*, would not have been able to resist showing him Calgary's version.

Chapter Seven

Heaven only knows what the objects of his studies are.
But here we are, and you must form your own
impressions of him.
— *A Study in Scarlet*

FOR MOST OF HIS LIFE, Dr. Earle Scarlett was, first and foremost a physician. Oddly enough, he seemed to think that he could build walls around various parts of his life and leave medicine out of some of those compartments.

"From the time I started into medical practice," he writes in an unpublished work after his wife's death, "I resolved that the world of medicine and my home and private life be kept strictly in separate compartments. All medical matters would be dropped as I entered my front gate. This custom I followed even more sternly as the years went by. Indeed I think my children often wondered what I did all day, and often far into the night."

He believed this of his children, although it was patently untrue, as they have attested; yet he did try to ensure that his entire life was focused on more than one discipline. He admitted to himself that Jean, his "perfect physician's wife," had always managed to bridge those different compartments.

His children's picture of him is that of a man sitting in his study, puffing on a pipe of John Cotton's tobacco, and reading. It's a true picture; he probably spent as much time in that study as he did on the hospital wards. There, he who boasted of having "printers ink in his veins" did his writing and his reading, and there he prepared "The Ram's Horn," an unpublished work which, considering the labour he put into it, says a great deal about his love for the printed word and his fidelity to Captain Cuttle's maxim, "When found, make note of." Many of those notes ended up in his commonplace books. Many ended up in this anthology.

Scarlett began writing the "The Ram's Horn" in 1944. It was, on the surface, an anthology of some of his favourite bits and pieces from literature, but in reality it was much more. The manuscript comes to almost 600 pages of single-spaced typing, and its quotations run from serious prose to comic verse, all larded with Scarlett's own comments. It is a literary eclectic circus, complete with sideshows and performers, with Scarlett as ringmaster, a document which could be called a diary, not of events but of thoughts. His son, Robert, once thought of having the manuscript published, and it's not impossible Scarlett himself may have toyed with the idea, although there doesn't seem to be any evidence for this. He seems to have intended it for his private use; if so, it is an extraordinary document. It reads as though the writer expected it to go before a much larger audience. Who, for his own enjoyment, would type out a list of lines from Christopher Fry's plays and then preface them with the following remarks?

> Writers, poets and dramatists of today are, for the most part, too cerebral, and their language flat and factual. You look in vain for any neo-Elizabethan colour and swaggers in their speech. Sean O'Casey is an exception. And Christopher Fry has shown a happy wit, an ability to handle amusing bombast, a cunning way with words, and, at times, a gift of rare poetry. I cannot understand why, after being acclaimed on the appearance of *The Lady's Not For Burning* he almost completely dropped out of sight. Probably the air of the modern Theatres of the Absurd is much too chilly for his romantic irony. His conceits are too sweet to be read rather than played.

Would someone write this for his own amusement? Obviously Scarlett did. One reason may have been that he was, by his own admission, a "random reader," and there is nothing more random than a wide-ranging anthology. As he writes in the manuscript's introduction, "There is no such thing as the perfect anthology – except the one that you make for yourself."

In this introduction, he looks back to his boyhood and remembers spotting a theological book belonging to his father emblazoned with a ram's horn trumpet. The trumpet was, to him, something of a symbol of what he sought in literature, not a tiny or shrill voice but a loud and strong note in his life.

This book is an attempt to recapture some of the echoes of the Horn. In ear-lier years it was always blowing, and if it blows less frequently today, I sus-pect that it is because with my contemporaries I am, like the Athenians, too prone to "Run about to tell, or to hear some new thing . . ."

I think most of us could die happily if we thought that on the other side the sounding trumpets would be a gathering up into great strains all the beauty and truth and human kindliness which came to us in single phrases and echoes at long intervals during our life in this world.

He also used the book as a way, once again, of stating that medicine is not be the be-all and end-all of a physician's life.

I suspect that this pastime of anthologising represents an attempt to indulge the pleasures of idleness, infrequent periods of playing truant from medi-cine. It may be that this attitude springs from a passionate, if well sup-pressed, desire for intellectual freedom, at least freedom from the more rigid categories of medicine, not from lack of interest in medicine. Only the phy-sician knows how jealous a mistress medicine can be. Certainly were it not for the consideration of bread and butter, and the jam on it, I sometimes think that I would gladly retire from the drive of professional practice tomor-row. I should then not be obliged to take a virtually all-consuming part in the job, but I should not cease to have a lively interest in medicine.

In 1966, twenty-five years after he started "The Ram's Horn," he re-viewed the anthology and found it still reflected the man he believed he was.

After having used these pages as a constant source of reference and refresh-ment for so many years I have recently been looking over the anthology criti-cally as a whole. It was begun more than 25 years ago. Now I review the work of my hands as one who is still some way, I hope, from making the last voyage in Charon's boat. And to my great interest and delight I find that there is very little I would change or modify. Presumably this means I have not changed very much. I do note that one implication in these jottings from a lifetime of reading experience is my firm belief that our ultimate aim in this world is to live in the knowlege of the absolute values — Truth, Goodness and Beauty. This I still affirm, probably more insistently than ever. So it is, then, that I can place my seal of approval on this "collection." The old man thus salutes the younger – and is content.

He wrote those words on his seventieth birthday. "I may say, in the words of my old mentor Thomas Browne: 'In seventy years a man may have a deep gust of the world, know what it is, what it can afford, and what 'tis to have been a man.'"

At least ten years after that he added a few more words to the manuscript. He was leaving "The Ram's Horn" to his family as a sort of memorial to himself and the wife he missed so terribly. His words further explain the genesis of the manuscript.

> This book celebrates my playing truant from medicine and my indulgences in the pleasures of idleness. I bear witness to the fact that what we call "the humanities" in medicine will not disappear as long as there is one physician left who has not bowed the knee to Baal or continues to hold the ramparts against the Philistines.

"The Ram's Horn" not only says something about Scarlett's thoughts but, with the commonplace books and his extensive reading, once again illustrates how he was able to turn out so many erudite columns. He was the consummate anthologist, collecting not just books, but thoughts. His method of working is illustrated by a glance at one of the books in his library which made it to a "preferred" list in "The Ram's Horn." This is the journal of the Swiss intellectual Henri-Frederic Amiel. The back endpaper and flyleaf are covered with page numbers, some from a 1978 perusal, some earlier, and the pages indicated are heavily marked in pencil or ink. The folded front page from *The Times Literary Supplement* for 7 March 1936 is inserted in the book. It contains an account of the various editions of Amiel's journals. The page, of course, is also underlined.

To some bibliophiles a book is to be treated with sacramental care. To Scarlett a book was a repository of fact and thought, and he wanted these at his fingertips.

Earle Scarlett did, in Thomas Browne's words, "have a deep gust for life." He was a person who loved to be, to use his term, "delighted" by the world. He loved good talk, good music, good books and good physicians. He even enjoyed good metaphysicians. He did not like idle chatter. Robert remembers that the only time his father was at a loss for words was when he was caught at some social gathering marked by "cocktail party" conversation. It was on these occasions that he picked up a reputation for aloofness among his peers.

Good books and good conversation could be found within his own walls, but while he enjoyed the security and warmth of the home he and Jean had established, he also loved to travel. Jaunts around North America to attend medical meetings were, of course, common during his days in active practice, but he and his wife mostly enjoyed visiting Europe and travelled there whenever possible. These trips increased during the 1950s, when the pressures of his medical practice were easing up and his job as chancellor of the University of Alberta left him more free time.

The trips revealed his romantic nature, not merely because of the places he and Jean visited but because of who, or rather what, went with them. Katherine, like many children, had a special friend, an orange-colored stuffed bear named "Orangee." In the 1950s she decided they should take Orangee with them and show him the world. Many parents might have balked at the idea of hauling a toy bear around with them, but Earle and Jean were delighted to do so. Naturally the bear had to have a passport, so it travelled on its first trip, to Paris, with a document devised by Jean and the children, issued by "The International Travel Agency for Small and Companionable Animals, Inc.," at Toad Hall in Calgary. As the family travelled about, Orangee's head would be protruding from Jean's travelling bag, giving him a good view.

"A family history is made up of all sorts of things – gay and grave, large and small," Earle noted. "Orangee's 'passport' was one of those things which seem trivial but which assume great proportions with the passing of years."

Many of their trips involved visiting British literary landmarks, such as the homes of Jane Austen and Thomas Hardy. In the late 1950s, with the pressures of practice lessened, the two made longer trips, especially to the Mediterranean. In a sense, Earle wrote in one of his unpublished manuscripts, they became "Mediterranean folk."

As Jeanie and I looked back over our various sessions of living in Greece, Spain, Italy and other places, and pondered on how they affected us, we came to feel quite clearly, I think, that our whole attitude towards life had changed. The rhythm of Mediterranean life is basically assured and contemplative. In Northern regions, such as our own, one protests about life; in the Mediterranean one accepts it ("This too must be endured.") In the blue clarity of sea and sky, life basically relates to happiness, not to ambition and "getting on."

At the same time, this romantic attitude, which appealed to that side of his nature, fought with his Victorian ideals, with something of what used to be called "the Protestant work ethic." He remembered one beautiful evening on the Island of Rhodes when he and Jean encountered a middle-aged Englishman on one of their walks. This gentleman asked how long they planned to stay. They would be leaving, they told him, in about six weeks. "Don't stay any longer," he warned, "or you will never leave."

It was advice Scarlett thought about, and with which his Victorian side agreed. The Mediterranean, he concluded, had much to offer but only if you left before its pull became too great.

To sink into acceptance of the Mediterranean is to invite dreamy acceptance of existence which leads to decadence, a loss of will and virtually to become lotus-eaters. We did not stay on the glorious legendary shores of the Mediterranean. We returned home refreshed and more fitted for life – after a searching and rich experience. But the flavour still remained on our tongues and continued to be savoured down the years. So in that sense we are Mediterranean folk. We are not the same persons we were before we set foot in Greece in 1959. We are infinitely better people who came to see new beauty and wisdom and finer perspectives in this troubled world.

For someone who enjoyed the Romantic poets with their classical allusions, the worlds of Italy and especially Greece held an endless fascination for Scarlett. Greece also had what was to him a physician's sacred shrine, the Island of Cos. But more than that, in Greece he found what he also valued: stability and continuity.

Dr. Peter Cruse is a professor of surgery at the University of Calgary whose interest in medical history was so strong that he volunteered to teach the subject at the medical school in 1972. The popularity of Dr. Cruse's course and the enthusiasm of his students gave Dr. Scarlett reason to believe that the love of medical history in Calgary had not drowned in a sea of materialism.

Cruse believes that the founding of the Associate Clinic's Historical Society and its *Historical Bulletin* stand as Scarlett's great accomplishments. He became a friend of the older doctor because they shared an interest in medical history and a love for Greece. They disagreed, however, over the matter of supposed Greek continuity. Scarlett took the line that the Greek peasants of today were the direct descendants of the Greeks of classical times. Cruse, with a better knowledge of Greek history, noted the many interruptions in the country's history. Cruse admits, with affection, that Scarlett could not be convinced this continuity had been broken – it was less a fact than an article of faith.

Scarlett and Jean's greatest trip began in 1959 and lasted fifteen months. To him this trip symbolized a transition in his life.

> After the ceremonies were concluded in which I was made a Chancellor Emeritus by the University of Alberta, stung by the realization that I was to be such an unholy "ancient of days," my wife and I felt the time . . . had come to act. So I walked out of our home here and, taking my Penelope with me to obviate any difficulties with possible suitors, we sailed directly for Greece and an extended Odyssey, the rule of the adventure being to wander as the spirit moved us, to avoid the tourist "beat," to live as much as possible with the people to capture as nearly as possible the genius loci in whatever place our inclination led us.

Scarlett began the trip by spending several weeks at the American School of Classical Studies, an enterprise with which he had been associated for years. The school treated him as minor royalty, something he enjoyed to the full. He and Jean followed this with a month in the Peloponnesus, visits to Cos, three months in Italy, a month in Yugoslavia, a month in Vienna, the four last months of the winter at a villa in Spain, and a tour of Ireland.

He also enjoyed telling of "our most remarkable adventure," which took place during an earlier vacation, in 1957. It may have been the first time Jean Scarlett, who distrusted airplanes, enjoyed her flight.

The four-month trip was to revolve around the Mediterranean. They were to return home in July, about the same time CP Air had planned to inaugurate a direct Lisbon–Montreal flight. So why not, the travel agent suggested, end the trip in Lisbon and take the direct flight to Montreal? The suggestion sounded reasonable, so off they went, their return flight secured, to tour for four months. They arrived in Lisbon four days before the flight was to leave. It had been a wonderful holiday.

> The following day I set about confirming our reservation for the flight to Montreal. I was speedily dismayed to find that nothing was known of such a flight. For the next two days I explored every channel without success. No trace whatsoever of CP Air either at Lisbon city offices or at the airport; the authorities knew nothing of such a project; the several counters at the airport

– British, Air Canada etc. – could give us no help. Cook's, usually omnipotent, threw up their hands; there were no posters or data at the Tourist Information Office. We were stranded.

The next day Earle returned to the airport "with a feeling of deep wonder and despair in my heart and a prayer to Hermes" on his lips. Searching the reception area and about to give up, he finally noticed a small Canadian flag on the British Airways counter. "Why the flag?" he asked the woman behind the counter. "In case anybody is interested in CP's inaugural flight," she replied. Within minutes he had his boarding passes. They presented themselves at the airport, checked their luggage, and were taken to a departure area where they waited alone. They watched with increasing wonder from the still-empty room as the plane was prepared. A red carpet was rolled out, a band began to play, and officials, including the mayor of Lisbon in his robes, assembled by the steps. CP staff finally joined them in the reception area to explain that due to problems in advertising and promotion the inaugural flight was to have been cancelled, but since CP needed the plane in Canada, it had been decided to go ahead with the ceremonies. The Scarletts were to be the only passengers.

The flight, needless to say, was "one huge party." The food was excellent – "no customary airline cellophane food." When the party flagged, the stewardesses set up stretchers, hung sheets for privacy, and the Scarletts slept the rest of the way to Montreal.

Earle could talk endlessly of the wonders he had seen, of the archaeological digs he had witnessed, of the continuity he believed bridged the ages, but it was this story which delighted him most. For once, the ingenuity of man and the wonders of modern technology had failed – but in his favour.

Earle Scarlett, B.A., University of Manitoba, ca. 1914.

Private Earle Scarlett, 4th Cycle Corps, with his father, Rev. Capt. R.A. Scarlett, 101st Battalion, World War I.

Jean Odell in Cobourg, Ontario, 1920.

HISTORICAL BULLETIN

Notes and Abstracts Dealing with Medical History

Issued Quarterly by the Calgary Associate Clinic
Calgary, Alberta

Vol. 1	MAY, 1936	No. 1

THIS number is the introduction to a series of articles which the Calgary Associate Clinic will issue quarterly to the members of the profession in the Calgary and adjoining medical districts. Each succeeding number will comprise a brief resumé of the proceedings of "Historical Nights," held monthly under the auspices of the medical staff of the Clinic, and will endeavor to review in turn the lives and works of that galaxy of Pioneers in Medicine, who, by their foresight, industry, determination and self-sacrifice, have established during the centuries the eternal principles of the Science of Medicine, and bequeathed to this modern generation the great historical doctrines on which those principles are based.

Jean Scarlett.

Scarlett outside his cabin near Johnson's Canyon, October 1954.

Top: Scarlett's beloved "Number 6," Mt. Temple Chalets, Lake Louise, Alberta.
Above: Earle and Jean hiking in the Rockies, 1956.

THE MEDICAL JACKDAW

A Doctor Reflects on His Critics

Like many another doctor in these days the writer of this causerie is often prompted to sit down and wonder just what the reformers and the earnest purveyors of advice would do to the medical profession if they had their way. The nature and magnitude of what would come to pass surpass the understanding of saints and sinners alike.

The individual doctor would seem to be pretty well liked, if regarded as a bit stuffy, narrow and too prosperous for the good of his own soul. But it is medicine as an institution which is under attack—and savage attack at that. The button is off the foils. There is no question at all but that a new era in the criticism of medicine and the doctor has come to pass. The stresses an⌐ ' ⌐ds of a changing social order ⌐' ⌐ ⌐s of political mat⌐· ⌐ which goes far ⌐ ⌐g at the

Scarlett's column in *Group Practice*.

Chapter Eight

Problems may be solved in the study
which have baffled all those who have sought
a solution by the aid of their senses.
— "The Five Orange Pips"

Scarlett the chancellor of the University of Alberta, no less than
Scarlett the physician, liked things done properly. In a sense he saw himself as a Roman senator charged with ensuring that the ancient virtues
were maintained and standards were upheld.

He fulfilled this role both on and off the Edmonton campus. According
to Grant MacEwan, mayor of Calgary during Scarlett's tenure at the university, a city alderman who was not too well educated but who had made
money in a paramedical area demanded that he be given an honorary degree. MacEwan was dubious but agreed to approach Scarlett. The chancellor replied, "He'll get one the day they award a degree for giving enemas."

Having "officially" retired from the medical scene and having taken
his Greek odyssey, he might have settled down to his writing, his clipping
and his pipes, and to playing the grand old physician to a younger generation. Some believed that was his appropriate position. A Quebec medical
publication called him the "doyen du corps médical à la clinique de
Calgary," a phrase which, he said, made him sound "like one of Napole-
on's marshals." But retirement was not to be. He wrote to his friend Bill
Bean in August 1961, "I am beginning to feel medical pressures crowding
in on my eclectic way of life. Just now I am occupied with sessions of acting as chairman of a small government commission to review nursing edu-

cation (that chameleon!) and serving as the medical representative on the board just starting to create a new 850-bed hospital."

Medical pressures had begun crowding him in more ways than one. Always a healthy person, he was beginning to discover some of the ills the flesh is heir to. In 1961 he contracted what was thought at first to be a "virus infection," that wonderful phrase which covers a multitude of ills, but which turned out to be pyelonephritis. "This," he joked to Dr. Bean, "brought me for the first time within the horrible precincts of my genitourinary colleagues whom I have always regarded with the deepest distrust. It was a near thing. However I escaped the ignominy and senile stigma of falling within that awful category of mortality – the 'prostate' community, thank God. It appears that on the counts of teeth and prostate I still have the blessed assurance of youth."

Not only was he almost as busy as before his "retirement" (he was now writing more than ever before), but he found himself plunged into several medical controversies, the first having to do with the new hospital. Calgary certainly needed one but there was less than general agreement over both its location and the kind of hospital it should be.

It had been known since the early 1950s that Calgary's rapid growth would compel the city to have a new hospital. By the late 1950s there was a chronic shortage of beds. Calgarians, medical and lay alike, were relieved when the government finally agreed to build. The first problem, though, was where to put the hospital.

Both city council and the medical community agreed that since the main growth of Calgary at that time was to the south, the hospital should be located south of the city's natural north-south dividing line, the Bow River. Money for hospitals, however, comes from the provincial government, and the provincial health minister, J. Donovan Ross, M.D., wanted it to be situated on what was then a relatively inaccessible hillside on the bald prairie in the north end of the city. He also wanted it to be a particular type of hospital.

According to Dr. Gardner, Ross himself had applied as a "family practitioner" for the right to admit patients to the University Hospital in Edmonton. Family medicine had yet to achieve the status of a specialty and his request was refused. The refusal affected some of his thinking towards the new Calgary hospital. Gardner knew the minister fairly well, so he wasn't too surprised when Ross appeared in Calgary one day and said, "I want to show you where the new hospital will be."

Gardner, shown the location, wasn't enthusiastic because it was so far from the centre of the city, a good twenty-minute drive from the Colonel Belcher Hospital, where the surgeon spent most of his time.

"Ah," Ross said, "that's nothing." He then outlined his ideas to the surgeon. "This is to be a family medicine hospital," he said, adding, "I could get on staff here. We want to have family medicine practised out of the hospital, not just out of an office or house calls. That's my concept. If ever we have a faculty of medicine here, this hospital is not to be the 'university hospital.'"

The location was not popular. As a *Calgary Herald* editorial pointed out on 12 November 1958, "It must be conceded that the new hospital will be a splendid looking edifice on this commanding and elevated site . . . but to the ambulance drivers, the harried physicians, the visitors and the patients themselves who must make the long trek, other considerations will be uppermost in mind."

Some Calgary doctors were also sceptical about the health minister's "concept," and there was a certain suspicion that the board, when appointed, would be without a representative from the medical community. As it turned out, that suspicion was unfounded. When the hospital's board was set up by order-in-council in July of 1959, there was one doctor named to it: E.P. Scarlett.

It was a logical choice in many ways. As Gardner notes, "Scarlett was considered, at that time, sort of the guardian of medicine in Calgary, so that if there was to be a new hospital he should have some say in the medical aspect of it." It was also a happy choice for Ross, since Scarlett agreed with him that family medicine should be practised out of the new hospital. Foothills Hospital thus achieved some of Ross's aims, becoming a centre for the study and teaching of family practice in Canada, a place where physicians could take a three-year postgraduate course in family medicine. It also, however, became the university hospital and home base for the University of Calgary Medical School.

Scarlett no doubt would have preferred the hospital to have had an easy birth, but that was not to be. It was conceived in controversy and its gestation seemed interminable. The sod was turned for the new hospital on 10 September 1960, with promises by the government that it would be in operation within three years. Although the delays were not Scarlett's fault, it was 1966 before Foothills finally opened, after a six-year period of labour problems, delays by the contractors due to dust storms, accusations of political intrigue in hiring practices, and debate about the hospital's place in a hoped-for medical school.

When it finally opened, the plaque dedicating the hospital carried words written by Scarlett – over which he had agonized for months:

> Within these walls life begins and ends. Here are reverence for life, a sense of the dignity of man, the distilled medical and scientific wisdom of years and a shelter from the winds of illness.

How many hospitals have an inscription at the entrance which says, without flinching, "People die in this building"?

There are, in fact, two inscriptions, both by Scarlett. The other, on the outside by the main entrance, declares that the building is dedicated to medicine, "a fellowship as old as mankind." Unfortunately, by the time the hospital opened, Scarlett had himself fallen into the arms of the practitioners of that ancient fellowship.

As a board member, Scarlett found the battles tiring. And in the midst of it he found himself chairman of a committee struck by the province to study nursing in Alberta. "It seems to me," he wrote Bean in May 1962, "that much of the agitation for nursing and indeed medical reform is not too unlike the old patent medicine vendor's line with much of the same hocus-pocus." Scarlett had a great deal of respect for nurses, but it was the respect of an old-time doctor. He understood that the demands of modern technology meant nurses' training had to be improved, but in the end his diagnosis and recommendations on this problem called not for surgical but for conservative treatment. The study had been promoted by a fear of a nursing shortage, much of it raised by nurses themselves, who believed their profession should have a higher status and that nurses needed more formal education. When "The Scarlett Committee" issued its report in 1963, it reflected its chairman's own desire to seek a middle path. It pleased no one.

The report stated there would be no nursing shortage if the government took immediate action to increase the amount of money spent on nursing education. It allowed that the University of Calgary should establish a course in nursing administration and that it might even set up other postgraduate nursing courses. But in general it reflected the view that nurses are best trained in the old-fashioned way, in a hospital where classroom study could be combined with a daily dose of working with actual patients. If nurses insisted on formal university training, the report stated, this should be combined with practical hospital work.

"Nursing as a whole," Dr. Scarlett told the *Calgary Herald* in October 1963, "is in vigorous health. There is not as much wrong with the profession as you would suppose from all the hue and cry which is raised from time to time. The fact is that nurses are just more self-conscious, more self-examining and more conscientious than many other groups."

Scarlett probably thought he was being complimentary, but although nursing organizations agreed with some of the committee's findings, they strongly condemned its support for traditional nursing education. They saw nursing as an academic specialty with some clinical work, and as this was becoming the norm throughout North America, they won their case. Within a few years Calgary's nursing residences would close down forever.

Scarlett was relieved when Foothills finally opened, and relieved too when the nursing report was handed in, even though he was not in favour of the final outcome. He would probably have agreed with much that his old colleague Dr. McNeil said about modern nurses.

> The relationship between nurses and doctors has all changed. They're often not supportive and co-operation is lost. The nurses are openly disrespectful and critical of doctors. There is a feeling among doctors that if you can employ one of the old hospital-trained nurses, you will have much better services in your office.[1]

By the time Foothills finally opened, Ross was in favour of a medical school in Calgary – but on the university campus and not alongside the hospital.

An ersatz Calgary campus of the University of Alberta had existed for years as an adjunct of the Alberta Institute of Technology, now the Southern Alberta Institute of Technology. The Calgary campus used part of the "Tech's" main building, a rather pleasant institutional-gothic brick structure, and some of the military huts left over from World War II that made up much of the Tech campus. Dr. Scarlett, as chancellor, seems to have acted as the unofficial campus doctor. Certainly some of those who attended classes were steered by campus officials to Scarlett's clinic.

In 1960 a proper campus came into being on the bald prairie on the city's northwest border, a couple of low, ugly, flat-topped buildings, referred to by students as "the button factory." These buildings, first known as UAC – University of Alberta/Calgary – was about half a mile from the site proposed for Foothills Hospital.

The possibility of building a medical school in Calgary arose while Scarlett was still chancellor and long before Foothills was opened. Dr. Gardner discussed the subject with Scarlett at that time and discovered the older doctor was dubious about the idea. He was afraid that the province could not support two schools and suggested that, instead, the existing medical school on the Edmonton campus be enlarged.

"Two medical schools in this province?" Scarlett said to Gardner. "No."

"There's lots of money and lots of students," Gardner said.

"You won't get the students."

"You will. There's lots of students being turned down in Edmonton and they have to go to B.C. or Saskatchewan."

"All right, then," Scarlett said, "but only as a part of the University of Alberta – as a subsidiary of the medical school in Edmonton."

"That might work in literature but not in medicine," said Gardner.

This was not the first time Scarlett had had to be convinced of the worth of a southern campus. Years before, when the idea arose of teaching university-level courses in literature and related subjects in Calgary, he had been less than enthusiastic. He wasn't sure whether the proper respect for literature could be inculcated in students studying at either the Tech or a local junior college, Mount Royal.

Scarlett came to agree that a medical school in Calgary might be feasible, so long as that did not mean watering down Edmonton's school or having a makeshift second operation. At the same time, when it was decided to mate the school with Foothills Hospital, Scarlett fought to ensure that Foothills' independence would not be compromised. He was reasonably successful. Foothills and the associated Tom Baker Cancer Clinic control the ground the school sits on, and it cannot expand without the hospital's approval.

When the time came to approach the University of Calgary about opening a medical school, those involved went first to Dr. Scarlett. Dr. Donald McNeil, president of the board of directors of the Alberta College of Physicians and Surgeons, had been a member of a committee struck in 1964 to discuss the founding of a medical school with Dr. Armstrong, president of what had by that time become a completely separate entity known as the

University of Calgary. McNeil was a friend of Armstrong's, but he and his committee (Drs. H.E. Duggan, G.S. Balfour, John Dawson and John Morgan) felt they should make one other call before approaching the university president.

> Before we went to see Dr. Armstrong we went to see Dr. Scarlett. We felt we wanted to talk to him. We wanted to act in propriety, to act wisely, and we thought we would lose nothing and would gain by going to talk to Dr. Scarlett. We went to Dr. Scarlett's home and sat down with him in his library. His advice was clear, and I think very worthwhile: "When you see the president don't do anything in a coercive manner, don't make any announcement to the papers in regards to the development of a medical school here, let it come from Dr. Armstrong. Don't put him in some awkward position by making an announcement prematurely in any manner."[2]

A third controversy turned into what Scarlett believed was a medical tragedy. After the war Britain's National Health Service came into being and influenced socialist governments around the world. The Canadian government watched as Tommy Douglas's Co-operative Commonwealth Federation (CCF) administration in Saskatchewan decided to implement a similar scheme. The idea of a national health plan in Canada was beginning to be bandied about, and a royal commission was set up to investigate the possibility.

Scarlett was not enthusiastic about the idea and spoke out publicly against it, suggesting that most people were already well served by physician-run plans. He estimated that only about 10 percent of Canadians could not afford private medical insurance and they were the only ones the government should worry about. He did not expect the public debate to turn into the vicious battle it did, culminating in Saskatchewan doctors going out on strike and the provincial government asking for a mandate to force doctors back to work.

Scarlett understood the profession's dislike for socialized medicine, but he could not understand how any doctor could refuse to treat the sick. The doctor's strike was a heavy-handed blow at everything he believed his profession stood for, every medical ideal he held. That the doctors had disobeyed the law made things even worse. To him obedience to the rule of law was the mark of a civilized society. In July 1962 he poured out his thoughts on the matter in a letter to his friend Bill Bean.

You doubtless noted that I have refrained from saying anything about the tragedy which is being played out in our neighboring province of Saskatchewan. I can barely bring myself to talk about it. Colossal mistakes have been made on both sides. But what concerns you and me is that the currency of medicine has been fearfully debased. At a time when we are fighting with our backs to the wall against a resurgence of barbarism and barbarians, it is depressing to find that the wave has swept within our own medical walls. In the process the most incredible hysteria has been bred. And no steadying voice has been heard from the leaders who are of the new generation behind me. Most oldsters in the ranks like myself have done what we could by way of counsel and writing, but we are brushed aside as sentimental old fools. I do not know of one of the old guard of medicine in this country or of the professorial and academic establishment who is not sick over the disaster and betrayal of medicine.

The fault, he believed, did not lie only with Saskatchewan doctors. Some of the fault lay with their American colleagues who had made out that prairie doctors were fighting a war against communism as well as socialized medicine. As he told Bill Bean:

I may whisper to you that not a little of the disease which has afflicted the profession in Saskatchewan and their subsequent incredibly irresponsible action is due to the virus introduced by the reactionary Chicago Nabobs of the AMA who more than two years ago, when the struggle with Premier Douglas and the government was joined, poured propaganda and fighting weapons into the Sask. medical forces and loaned a so-called public relations man or two to direct the battle. The result, as my newspaper friends tell me, was one of the most horrible and stupid campaigns ever conducted in Canada, with even the vulgar "plebs" turning against the profession and the government winning a resounding victory. But for some reason or other the deadly AMA virus has lasting results. And now has given rise to a course of action which is not open to any individual in a democracy – deliberately deciding to disobey a law of the province.
It is going to take the profession at least two generations to repair the mischief. It is nothing short of tragic that the rank-and-file are so blind to the trends and duties of medicine in a modern society, and have defiled the temple which has been built by medical men of good will everywhere, at the same time ignorantly repudiating the views and good offices of fac-

ulties of medicine, public health men and all thoughtful citizens. It seems
that the antibodies, viciously produced by the public media of opinion, to
Communism are almost as bad as the disease itself.

The 1960s, which had begun so well for Earle and Jean Scarlett, turned
out to be a decade of upheaval. Scarlett the man applauded when students
spoke out against materialism, but Scarlett the academic, and especially
Scarlett the Victorian, stood aghast as students took over campuses,
shouted down their professors, and disrupted classes. Even though Alberta universities remained immune to extreme student antics as occurred
at such institutions as Sir George Williams in Montreal and Rochdale in
Toronto, society itself seemed threatened. The doctors' strike had dis-
turbed him at the beginning of the decade, but there was worse to come.
Earle Scarlett, who had nothing but good to say about French Canadians,
was heartbroken when the FLQ (Front de libération du Québec) began setting bombs and shedding blood. He believed in the ultimate goodness of
man, but his strong faith was tested as Canada's bicultural struggle sank
into its nadir and terrorists murdered Pierre Laporte, a neighbour of his
daughter Katherine.

Chapter Nine

For goodness' sake let us have a quiet pipe and turn our
minds for a few hours to something more cheerful.
— "The Adventure of the Speckled Band"

In the late 1950s Earle Scarlett was "chancellor emeritus" of the University
of Alberta and "president emeritus" of the Associate Clinic: he also had
"emeritus" status at the Holy Cross Hospital. Technically, "emeritus" means
"honorably retired from active service," but in Scarlett's case, "honorably"
was the only part of the definition that was completely suitable. "Inactive"
was not part of Scarlett's vocabulary.

Behind him was a glorious jaunt to Greece; ahead were the Foothills
Hospital, his investigation into nursing education and numerous intrusions
on his time. He was busy, but not busy enough. He no longer had a regular
outlet for his literary criticism, his love for medical history, and his strong
feelings on the art and science of medicine.

In Europe, however, he had run into Ed Jordon, editor of *Group Practice*,
a familiar journal to those at the Associate and other clinics. Jordon suggested Scarlett might write a light, "medical reader" type of column for his
magazine.

The result, in the September 1960 edition, was a column headed "The
Medical Jackdaw," the heading done in the form of twigs and accompanied
by a drawing of a black bird – something of a cross between a crow and a
budgie – gazing fondly at a nest full of purloined objects.

Readers of his historical journal might have remembered the title from
an essay published in 1952, "The Jackdaw's Nest." They may even have been

familiar with Gerald Bullett's 1939 anthology, *The Jackdaw's Nest*. Only those who knew Scarlett well, however, would have had an inkling of how large this nest would be and how long this ageing bird would inhabit it.

Scarlett opened his column by establishing common ground with his readers and explaining the underlying purpose of what he said would be a "column of small talk." His experience as a doctor was within a medical genre they would understand, he noted, as he had thirty years' experience in a clinical setting behind him.

> I have proved the value of this kind of medical collaboration and know how the characteristic individualism of the medical man can be conserved and allowed to flourish to good effect in such an environment. I know that the average physician coming into the composite life of a medical group is at first like the Scotsman refusing to row in an eight-oared crew for fear of losing his individuality. It takes a bit of time for him to realize that by pulling in unison with the other members of the crew he can cover the course much more efficiently than if he paddled along on his own.[1]

He went on to "remark in passing" that from the vantage point of a clinic member he had "come to realize that many of the worst ills of medicine today are due to two tendencies in modern society – excessive centralized administration and legislation related to the concept of the welfare state, and the rapid growth of science and technology. As a result, the chief concern of medical practice – the individual human being – tends to be forgotten and a host of disorders spring up which continue to plague us."[2]

Having delivered this passing remark, he suggested the "Jackdaw" column should be seen as a casual conversation between Scarlett himself, a "lively member of the *ancien regime*," and the reader of the journal. At times the Jackdaw would be fanciful but, he explained, it could also speak in a serious tone.

It would indeed be both fanciful and serious but this curious bird would have little to say in the next few years about the clinical experience.

With the experiences of his Greek sojourn still fresh in mind, Scarlet not surprisingly chose to let the Jackdaw, following his introduction, talk about Greece, specifically Scarlett's beloved Island of Cos, which he reasonably enough considered sacred to men of medicine. Although the initial column included a few shorter items, as was usually his practice, it attempted to bring his readers back to their roots. He admitted that the scientific side of medicine may be able to get along without its history, but he added, "Medicine as an art cannot afford to neglect the great statements of principles and aims which have provided its life and motive power down through the centuries. Medicine must contribute more to man than antibiotics and vitamins and surgical finesse."[3]

After hitting his readers with a bit of philosophy as a portent of things to come, he concluded, in his usual style, with a delightful pastiche of Macaulay's "Horatius," a "Lay of Modern Medicine," in which Lars Porsena becomes Staphylococcus Aureus. Given that he took it from the 1909 edition of *St. Bartholomew's Hospital Journal*, it gave readers some idea of just how wide-ranging this "conversation" was to be.

The fourth bauble in the Jackdaw's nest contained a short piece on doctors' wives, a subject he would enlarge upon later here and in the *Archives of Internal Medicine*. Having whetted readers' interest in that subject, Scarlett passed on to the strange titles of medical papers and an old "monkish" litany in both Latin and English: "From twice-cooked food / from an ignorant doctor / from a reconciled enemy / from a wicked woman: Good Lord deliver us."

"Probably written by one who had 'suffered many things of many physicians,'" Scarlett commented dryly.[4]

He encouraged doctors to take their profession but not themselves too seriously. When necessary he defended the profession, even against what might seem to be a small point. In the February 1961 issue, he rails against an unusual target: *Harper's* magazine. The magazine had published, in its October 1960 issue, an article exploring the problems of medicine. Scarlett did not argue with the editorial content but he did take issue with the cover illustration, a winged staff with two serpents wrapped around it. (This confusion in symbols is something Scarlett would encounter again a decade later when the Canadian Post Office issued a commemorative stamp honouring Osler.) The Jackdaw was outraged not just because the twin-serpent staff of Hermes had been confused with the single-serpent staff of Aesculapius (a common error), but because Hermes was the god of thieves and his staff was associated with commerce: "To have this emblem foisted upon the medical profession is a most unhappy, if not intolerable blunder, especially when we note that in the ancient statues of the god he is shown not only bearing the two-serpent staff but also carrying in his left hand a filled purse."[5]

Given Scarlett's own approach to life and literature, it's not surprising that nine months later he followed this attack on misinformation with a column on serpent worship and the serpent cults of Greece. Scarlett always liked to make sure his readers received a well-rounded meal.

Some of his writing from the early 1960s seems very modern indeed: the problems of working with an increasingly ageing population – today's ageing baby-boomers were still in school – and the rise of paramedical specialtes. He also worried, in print, about the decline of his profession and the other "learned" professions, all of which he regarded as a barrier against state tyranny. The mark of a profession was a combination of personal independence and a sense of public service, both founded on corporate discipline. This was the time when socialized medicine was on the march in Canada, and Scarlett was of two minds about it. It would harm the profession, he believed, but the profession had only itself to blame. The suggestion that doctors might attempt to fight socialized medicine by withdrawing their services horrified him. That way lay social anarchy.

> Within the profession the situation has been worsened by the narrowing effect of specialization and the absence in our ranks of a large number of men with what may be called a liberal education and an informed outlook on society and the world. There is an increasing difficulty in holding that minimum of freedom necessary to maintain the integrity of the profession. And the contagion of materialism and the worship of "the bitch-goddess, Success," have made heavy inroads, leading physicians to adopt practices and ends out of line with any body which calls itself a guild or a historical order in society.[6]

After a serious homily he would beguile his readers with light verse, including clean limericks, epitaphs, proverbs and his favourite Rabelaisian curse: "May Saint Anthony's fire scorch their snouts." He would discuss famous people who preferred to spend their lives in a prone position in bed or on a sofa – Florence Nightingale and Charles Darwin to name but two – or go on at length about 1961 being the duo-centenary of Leopold von Auenbrugger's *Inventum Novum ex Percussione Thoracis Humani*, which, in 1761, introduced the technique of chest percussion as a diagnostic tool. He introduced his readers to unusual cures, such as Samuel Pepys's practice of carrying a hare's foot as a specific against flatulence.

In November 1961 Scarlett wrote one column which doubtless gladdened the heart of Hospital Minister Ross. Scarlett was well aware that the idea of family practice as a specialty was the cause of a good many snide remarks inside the profession. Doctors argued that a family practitioner was simply a general practitioner with a fancy title. Scarlett joined the fray by announcing in his column that he disliked the term "general practitioner," since it suggested a GP was a medical jack-of-all-trades.

> I prefer the term "family doctor." There is much cloudy thinking and dou-
> ble-talk, and there are too many platitudes poured out into the discussions
> of this subject . . . But the paramount fact is that these family doctors — the
> yeomanry of medicine — constitute in my own country, Canada, at least 65
> per cent of the profession. It is high time that we got their place in the scheme
> of things into better perspective, that we stopped giving our main concern to
> research and the specialties, and set about organizing our training and prac-
> tice accordingly.[7]

It is doubtful that his word carried much weight in this argument, especially as it was printed in a journal with an extremely limited readership, but at least the protest came from a specialist rather than a GP.

Scarlett delighted in poking fun at people, including himself. When readers pointed out that "daw," as in jackdaw, meant "simpleton" or "sluggard" – "slattern" being the feminine – he printed it with glee. "Ah well," he concluded, "it is all very uncomplimentary, and I settled back in my nest quite chastened, taking what comfort I could from the battered but still (in most quarters) respectable adjective 'medical,' which I hoped might redeem the disreputable, trifling jackdaw himself."[8]

It did not take readers long to figure out that the jackdaw was a literary bird, but since he didn't always give his sources, they must have wondered how much he really read. In the January 1963 edition, he offered a definition of the word "clinic" which had caught his fancy: "Generally used to signify a quack." Where did he pick that up? Was it from his own 1730 copy of *The Lexicon Physico-Medicum*? He didn't say, but fans of the Jackdaw could be forgiven if they imagined the bird inhabited a nest of ancient, leather-bound volumes. (The book in question is not included in the Scarlett Collection in the University of Calgary medical archives. Given his eclectic book collection, who knows? His collection, which began to metastasize and take over the entire house after his children left home, was dispersed after his death on his orders.)

By the mid-1960s tobacco began to be viewed less as a social pastime than as a dangerous drug. Scarlett, surrounded by John Cotton fumes, couldn't resist improving on what his friend Dr. Johns had written on the subject in the *Historical Bulletin*. He praised the joys of a good smoke. He went through its history, noting that it had once been seen as a specific against syphilis and had eventually found a place in the pharmacopoea as a sedative, emetic, diuretic, narcotic, and cure for cancer. (He did not mention that as late as the 1930s some medical books aimed at laymen advised a smoke after meals to cleanse the mouth.)

> The practice of smoking is now more general than ever before. And what is the musing philosopher to think of it all? Answering the question amid the clouds of smoke from my pipe, I can only speculate that man must have some form of narcotic or stimulant, that in tobacco he finds a means of escape from the pressure of the time, that it has become a vehicle and symbol of friendship and social intercourse, and that for the true votary it is a minister of grace and a solace for the indignities of time and age.[9]

As the Jackdaw, he was erudite about tobacco. In person, his romantic nature took over. Opening a fresh tin of tobacco, he would say, was one of the small but lasting pleasures of life. He could not bear to throw away a pipe after it had outlived its usefulness. After his wife's death he would look at the racks of pipes in his study and call them his "dear old friends," "silent witnesses to years of bliss and contentment," which at times "seem to lean forward asking to be cradled again for a session in my hospitable palm."

In September of 1963 he turned to a favourite topic, the doctor's wife. Perhaps this was because, as far as he could find, few doctors had written about their wives. He noted there seemed to be a new variation in the boy-meets-girl theme of romantic literature, something he admitted he knew little about. Instead of romance in Venice or a foreign embassy in some steamy land, the new setting was the operating room. There doctor and nurse feel the first twitch of romantic attachment as, above their masks, their eyes meet. "So far the morgue and the anatomy room have not appeared as the scene of romance," he admitted, "but give our enterprising popular novelists time."[10]

Between praising "the doctor's handmaiden" and relating a humorous medical story, Scarlett would abruptly switch subjects and bring out some fascinating piece of medico-literary knowledge. In the October 1963 column, he discussed how doctors should respond to critics (he agreed with a friend who said, "Tell them to go hang!"). He debated, etymologically, the differences between "râles" and "rhonchi," and spiced this stew with quotes about owls, dogs and anatomical errors from popular writers. He then devoted an entire page to medical problems in Hamlet, explaining why Queen Gertrude would describe the lean and melancholy Dane as "fat."

The Jackdaw's nest was indeed full of an odd collection of gewgaws, throwaways and gems. Much of what he wrote appeared only slightly reworked in *Archives of Internal Medicine.*

It's unfortunate that *Group Practice* was not read by students of literature as well as practitioners of medicine. Scarlett continually found an excuse to enliven a medical article with illustrations from Shakespeare, Keats or other heroes of English literature. Some columns were totally devoted to literary aspects of medicine, as in the essay "A Medical Profile of Jane Austen," which appeared in the December 1969 issue.

In 1963, while the battle between doctors and the Saskatchewan government raged, he attended a conference on group practice. It left him puzzled and depressed. His own clinic had had its problems in the early days, but the principals had tried to give its members a feeling of cohesion to buoy up their sense of calling, which, he wrote, had deserted them. The administrators, he noted, complained that their main problems resulted from an absence of team spirit, a "chilling materialistic attitude on the part of most of the members, a concern for security in a narrow sense in which the drawing allowance, the hours of work, the chances of promotion were the paramount matters."[11] The veterans at the meeting disliked the trend, but they seemed to have given up even complaining. Writing about the subject, he admitted, would probably cause younger readers to wonder when the Jackdaw would get off his high horse and stop moralizing. Rather than doing so, he persisted, suggesting that technology should not be taught in medical schools at the expense of the humanities. Schools should emphasize medical ethics, which should also be discussed at medical meetings.

But some may ask, "Just what is this medicine you are talking about anyway? Isn't it just one more of the many occupations of society?" In my view it is very much more than that. As I have said so often in other places, medicine is a technique wedded to a reality which transcends it and makes it vital. The real core – the essential reality – of medicine is not an affair of clinics, laboratories, institutions. It is the meeting of doctor and patient when a person who is ill seeks the advice of a physician whom he trusts. All else in medicine is secondary to that. This is the relationship that must be fostered and preserved. This is the ancient and central rite of medicine demanding a standard of conduct and manners, understanding, sympathy, responsibility and knowledge. It is a relationship capable of infinite content and variants.[12]

It may have given Scarlett pleasure that the University of Calgary Medical School announced it would fill its student ranks not only from those who had specialized in the sciences, but also from students of history, literature and the other humanities.

The Jackdaw plied his colleagues with suggestions for reading material. In his January 1964 contribution, he devoted the entire column to a list of books he believed should be on every physician's bookshelf. Many were actually on medicine and some were fairly recent publications. Others were obvious favourites: Sir Thomas Browne's *Religio Medici*, *The Complete Sherlock Holmes*, the letters of Keats, and the Oxford books of English prose and poetry. And, of course, he included a number of books by Sir William Osler, Osler biographies, such as that of Harvey Cushing, and, to complete this historical chain, John F. Fulton's *The Life of Harvey Cushing*.

"To the doctor who says there is no time to devote to reading, the answer is what it has always been – one does not *have* time; one *makes* it."[13]

Occasionally, he came up with an unusual but telling and topical piece of information. After a visit to Ireland, he discussed the Irish sweepstakes. He had, he wrote in his April 1964 column, expected Irish hospitals to be in a fairly good financial position. He didn't realize the Irish government took 25 percent off the top as a type of tax on hospitals and that only about 20 percent of the take ended up in hospital coffers. He wrote the column because it had been proposed that sweepstakes could underwrite Canadian hospital costs, a proposal he thought total nonsense. His answer to this idea, advanced by politicians, reformers and journalists, was that if Canadian sweepstakes were created and brought in an amount equal to Ireland's during the last thirty-three years, they would keep Canadian hospitals going for approximately four days.

The run-down condition of Irish hospitals distressed him. To Scarlett, hospitals were where doctors and nurses learned the meaning of their professions. It was natural that, as a physician, he had an affinity for hospitals, but his words were those of a romantic. Who else but a die-hard romantic could write these words: "Did you ever look at a great hospital at night with the light streaming from its windows? It always reminds me of a great ocean liner lifting its great and strong outline above the roof-tops of the city, one of the proud, strong, secure things in this world."[14]

As time went on, *Group Practice* changed, taking on a larger and more modern format. However, the Jackdaw remained in his nest, philosophizing over the old values and trying to come to grips with new problems: the use of hallucinogenic drugs, the medico-social aspects of abortion (he applauded the liberalizing of the abortion laws), genetic research and, finally, pollution, which he regarded as humanity's greatest danger.

In July 1969, with his own health an issue, he puzzled over one of the worst dilemmas of medicine, the "right-to-die" legislation. He noted that an English doctor, Kenneth Vickery, had been the victim of a witch hunt after suggesting that vigorous methods of keeping alive a patient who might be described as a "vegetable" might be out of place and that allowing death with dignity might be preferable. It is obvious that Scarlett's sympathies were with Vickery, but he strongly disagreed with the corollary: legislation that gave a doctor permission to induce death. Doctors should be prepared to remind the public, he wrote, that the practice of euthanasia violates the fundamental principle of medicine – reverence for life.

> Doctors do not wish to play God nor induce death, no matter what the motive. Today we know as never before that we are all involved in mankind. One of the basic values of medicine and of civilized society is the respect in which human life is held. Anything which, in however small degree, diminishes the value of human life, diminishes the value and stature of our society.[15]

As the decade grew old, the Jackdaw lost some of its spirit. The magazine, *Group Practice,* he felt, no longer had much time for him, and though the nest remained filled until June 1971, in 1970 and 1971 the Jackdaw himself lacked bite. His last columns were generally devoted to oddities of the English language, with the occasional excursion into medical history. He had seen the end coming when Ed Jordan left as editor. He did not like the changes in the journal, and sad though it might have made him, he was content to let his column die.

As the Jackdaw, and as the "Doctor Out of Zebulun," Scarlett was something of a rare bird, though not quite Juvenal's *rara avis in terris nigroque simillima cycno* – he was not a black swan, medical or otherwise, despite his often bleak and pessimistic views on the future of medicine. His columns were unusual because of their style, if not unique in their range. Scarlett, born while Queen Victoria still reigned, lived to write in an age of jargon, of minimalism, of literary carelessness. He was a relic of an age which believed that a person daring to wield the pen of the writer should be as well versed in the fundamentals of language as a surgeon should be in the fundamentals of anatomy. He could appear wordy and even pompous, but he loved language. He believed that words should not only communicate ideas and feelings but, when properly chosen and set together, have a beauty of their own. An essay, therefore, was to him an exercise in beauty as well as truth.

Chapter Ten

One's ideas must be as broad as nature
if they are to interpret nature.
— *A Study in Scarlet*

READERS OF THE *Archives of Internal Medicine* opened their January 1962 edition and found that its new editor, William B. Bean, M.D., had something unusual for them. Bean wished to introduce "an old friend whom I have met only through the enchantment of the written word, one who in correspondence has been able to delight me with the benediction of his wisdom and the warmth of his friendship, as well as to provide for me a benchmark against which I can judge character and excellence."[1]

Over that introduction came three singular lines of type: a headline, "Doctor Out of Zebulun"; a quote explaining the heading, "Out of Zebulun they that handle the pen of the writer (Judges 5:14)"; and "Gleanings from the Commonplace Book of a Medical Reader."

Ten years before this, Dr. Bean explained, a newspaper in Des Moines, Iowa, had printed excerpts from a speech Scarlett had given in Toronto titled "The Physician's Role in a Modern World." What he read prompted him to send a letter to Scarlett, a letter which had turned into a friendship by correspondence. Now, he wrote, *Archives* readers were about to be exposed to a mind which had enriched his life and which he believed would do the same for them.

Here we see a scholar, physician, thinker and reader, a man of almost unnumbered distinctions. It is with more than a little pride and a great feeling of pleasure that I have induced him to illuminate some pages of the *Archives of Internal Medicine* in a monthly column, which he is pleased to label with a resounding biblical title . . . Though I shall not now receive so many pleasing letters from Dr. Scarlett, I am more than compensated by knowing that I can share with you the fresh and wise reflections of an old friend.[2]

The letter which began this long correspondence was a simple request for a reprint:

In our Des Moines paper I have read some excerpts from your address to the Canadian Medical Association and would like very much to have a copy of the entire paper if you have reprints available. Also, anything along similar lines would be most welcome. Thank you very much.[3]

Dr. Scarlett, as chancellor of the University of Alberta, was used to such requests. His formal reply was sent almost two months later and contained an apology for the delay and a promise that reprints would be sent as soon as possible.

The formality in their letters gradually dropped away as the correspondence continued. It became clear that both were carrying on a campaign to ensure that the science of medicine did not obscure its art, and that materialism did not undermine the discipline's essential humanity. It was, in fact, a mutual admiration society.

A few years later Scarlett would thank Bean for sending him reprints of the Iowa doctor's work: "As always they were a delight and an inspiration (if I may use such an over-worked word). I have come to regard you as possessing one of the most original, informed and literate minds in American Medicine."[4]

In 1961, thinking towards his coming term as chief editor of the *Archives*, Bean remembered the columns Scarlett was doing for *Group Practice* and decided he would like to see the "Medical Jackdaw" fly to another tree.

I have just read over your latest "Jackdaw" and find it delightful. I feel that it is being thrown to the wolves appearing as it does in the journal where it is buried, a journal I am ashamed to say I had never heard of before I found out about your column. What I want to ask you is this; I am taking over the *Archives of Internal Medicine* as chief editor on the first of January.

Wouldn't you like to let us have that for the *Archives*? The *Archives* now has a subscription list somewhere between 60,000 and 70,000 and is I think the most widely read of all journals of internal medicine, being outstripped only by the *J.A.M.A.* and the *New England Journal of Medicine* and perhaps by the *British Medical Journal* in England, though I'm not sure of this. It's certainly well ahead of the much more literate and delightful *Lancet*. I don't know what financial arrangements, if any, you have with the people that you contribute to now so you might let me know, but I should think there would be no trouble with any modest stipend.[5]

Scarlett felt honoured by Bean's offer but one aspect of it was unacceptable. *Group Practice* had asked first.

As far as "Jackdaw" is concerned, I cannot let down Ed Jordan, the editor, by taking the column away from *Group Practice*. The small audience bothers me not a bit. I like a small group anyway. Jordon sought me out in Europe, and I agreed to do a column for him which has become well established and seems to be well received. It simply would not be cricket for me to pull out the "Jackdaw" and put it in another nest . . . Jordon is a good and fair editor, a first rate fellow, a man of the craft, an all-round physician, and my relations with him and his little journal have been most happy.[6]

But he added:

At the same time I should dearly love to work with you. Under the circumstances I can think of only one alternative. And that is that you might care to have me run a column in the *Archives* of a somewhat similar character. Another title, of course (I could dream up an appropriate one, I think). Probably a broader frame of reference, a little different emphasis, even developing some themes in installment fashion etc. But generally speaking in the commonplace-book sort of vein.[7]

Bean approved of Scarlett's suggestion so long as he was able to keep up with the monthly deadlines. But what about a title?

Scarlett replied that deadlines kept the mind from falling into sloth and procrastination. As for the title, he suggested two: "Doctor Out of Zebulun," followed by the biblical passage; and the old title of his *Bulletin* column, "A

Medical Miscellany – From the Commonplace Book of a Medical Reader."
The chosen result, of course, was a combination of the two. As for the stipend, he would leave that up to Bill. "I have written for so long in the medical world for nothing beyond the glory of the medical gods, that a modest stipend is acceptable, if only to support the limited resources and modest requirements of a superannuated physician."[8]

The first manuscript was sent off a month later. Scarlett acknowledged Bean's kind words: "They will be perennial flowers along my path from now on. Please God I may continue to value them, and not do them discredit. This exercising one's mind with words is a perilous business."[9]

At the same time, Scarlett noted a few problems. He wanted the headline reset (it was), and he worried about requests for reprints, which had already begun to pour in. He simply could not afford to supply them and would have to say they were not available.

This first column, which followed Bean's introduction, was headed "By Way of Prologue." Both in style and content it warned readers what they could expect from the Zebulun doctor. In 1962 North America had yet to see the educational revolution in which proponents of self-expression would submit established linguistic standards to the firing squad. But even then Scarlett would have seemed a relic from the era of spade beards, frock coats, and wooden stethoscopes. Beginning a column without an introductory note, he wrote, is like entering a drawing room without making a bow. The following prologue, he said, would be his bow.

I welcome this assignment because it provides an opportunity to repay, in however an inadequate measure, the debt which I owe to the world of my time in which it has been my privilege, since the turmoil of World War I, to live as a freeman in the borough of Medicine which I like to think is still within the confines of the City of Light and Learning.

A preface is also in part an apology. Thus, while I may claim the merit of being an internist *emeritus*, I am bound to add that I recognize certain antedeluvian characteristics in myself – I wear a waistcoat, carry a "turnip" watch, am fond of oatmeal, smoke an Edinburgh tobacco, am devoted to Sherlock Holmes, and still use a venerable fountain pen which I fill periodically with an eyedropper. Also, for more than thirty years I have kept a commonplace book, a perpetual anthology of notes and reflections, which will form in a large part the staple and source of this causerie. Thus I stand before you as a veteran of wars and an old gladiator in the medical arena.

Queer characters these old soldiers! I recall how Aristophanes, speaking of the tough old survivors of Marathon who, in the way of old men, had the habit of going about Athens at unearthly hours in the morning, described them as having "Lights in their hands, old music on their lips; Wild honey and the East and liveliness."

What "wild honey" I shall dispense in this column remains to be seen.

For I am fully aware of the difficulties of providing medical and para-medical chitchat from a tiny stall in the busy medical forum. A distinguished member of our profession, Hieronymous Fabricius of Aquapendente, who more than three centuries ago was Professor of Anatomy and Surgery at the University of Padua and the teacher of William Harvey, the Englishman, on one occasion promised his audience a discourse *absolutissima, fructuoissima, luculentissima*. I am promising nothing of the sort, the more so because it is recorded that his listener went away from Fabricius's lecture muttering, *Imperfectissima, obscurissima*.

What I propose is, if possible, to create in this corner a sort of oasis of thinking, reflection and contemplation in quiet surroundings, a retreat where we may enjoy the simple liberty of leisure moments. In these days all of us – medical and lay alike – are "turmoiled by our Master, Time." The future is menacing and hidden. But bound by a common humanity we are called upon to build our defenses against the new barbarism, the corroding materialism, and the atom bomb mentality of fear and impious pride. Although this is an age of crisis, I very much doubt that it is "closing time in the gardens of the West." Nonetheless there is a bleak wind blowing from the steppes and, in our inner circles, a Freudian sirocco from Vienna.[10]

In this, his first column, he went on to quote Sir Walter Raleigh, Dr. Johnson, Pericles, and Prof. Jacques Barzun, and to give an account of Dr. William Harvey's attempt, and the subsequent lawsuit, to cure a man of kidney stones with a "secret remedy."

It says a good deal about the journal's readers that, instead of suggesting Scarlett belonged with the study of paleontology, they begged for reprints of his columns. He wasn't a universal hit, but a good number of readers did indeed appreciate having an "oasis of thinking, reflection and commentary."

Dr. Scarlett began by reflecting, or rather editorializing, on the English language:

Standing up in the medical forum, we boldly proclaim certain things. We cannot make too large a claim in the matter of language. On the quality of a nation's language depends to a great extent the quality of its life and thought. The relation is reciprocal. As E.M. Forster once said, "If prose decays, thought decays and all the finer roads of communication are broken." Language is still our most powerful and subtle means of communication. It can win souls, break hearts, rally a nation, rouse poetry in the heart of man, stir the worst emotions, break down the walls of prisons and raise them too. We should approach words with humility and reverence.[11]

The point of this was, to him, practical. "By and large," he wrote, "science is not a good friend of language," and doctors are too often guilty of being enemies of this common tongue. Proper use of language was a moral imperative. To read "the tidal wave of medical publications," he wrote, is to "contemplate medicine being drowned in its own secretions . . . It is infinitely worse if those secretions are of a heavy, tawdry, and third-rate nature."[12]

I remember that once many years ago when I criticized a medical student for his mode of expression in a certain examination paper, he said to me, "You got what I was driving at, didn't you?" When I answered, "Yes," he replied, "Then what difference does it make?" There is the issue stated baldly. There *is* a difference – the difference between professional integrity and careless, criminal opportunism.[13]

The third offering continued with Scarlett's thoughts on words, but having editorialized heavily on the need for good, colourful language, he now proceeded to have fun with the idea, giving his readers insight into the meanings and histories of words, especially those used by the medical profession. *Atherosclerosis*, he pointed out, literally means "hard porridge." Someone should have pointed out to Scarlett that this could be an uncited side effect of eating his favourite breakfast, oatmeal. Physicians who want a better term than "shaky," for weakness after an illness, might consider the Lancashire "wommacky," which means the same thing.

For his fourth column he delved into his files for letters from an old friend, Dr. W.W. "Billy" Francis, Osler librarian of McGill University, and wrote an essay on Osler's experiences in delivering babies.

During that first year he wrote about medical students, medical dress, nurses of the 1860s, and Dr. Edward Adrian Wilson, who died during the Scott Antarctic Expedition. He also wrote a seven-page essay on John Keats, using some material from a similar column in the *Bulletin* but adding to it. He wondered whether this second, more mature attempt devoted too much room to Keats. An apology accompanied the manuscript, but it was less than half-hearted.

> The only excuse – Keats has always been my chief passion (cultivated privately, as I have never found anyone to share my passion since dear Louis Holman, the Boston Keatsian Collector died – oh, yes, I should mention Christopher Morley). But I am always firm in getting Keats' connection with medicine properly delineated instead of the ill-informed rhapsodic mention which is usually the fashion . . . I could not shorten it, and somehow it seemed sacrilege to cut it in two. In future I shall stay within my prescribed limits.[14]

The Keats column resulted in a request for a reprint from Prof. Sir Charles Dodds of London. In this case, Scarlett asked his editor to supply it, since both of them knew the professor.

Meanwhile the letters were pouring in from as far afield as Budapest and Norway. Doctors wrote from Yale, from Cleveland, from New Haven, Connecticut, and from Burbank, California, to congratulate him on his columns, especially the two on language, and to give him examples of other unusual words or etymological lore. Some, of course, took issue with him. The interest shown in words delighted him, but he felt the need to write something of a disclaimer. As usual, he took it for granted that his readers were as well read as himself.

> In the midst of all the argument let me make one disclaimer. I had no intention of being pontifical or of laying down the law. Then and now I emphatically deny any thought of being a *super grammaticam*. I remember too vividly the example of the poor emperor at whom History has pointed the finger of scorn for more than 500 years. You will remember the story. At the Council of Constance in 1413 the Emperor Sigismund, opening the Council and speaking in Latin, used a feminine adjective with a noun which is neuter. A trembling ecclesiastic whispered to him, "Pardon, your Majesty, but *schisma* is of the neuter gender." Whereupon the Emperor loftily replied, "*Ego Imperator Romanus sum, et super grammaticam.*"[15]

For years Scarlett had been editor of the *Historical Bulletin*. Now he was up against his equal in erudition and one who, as chief editor of the *Archives*, did not let his columnist get away with anything. Scarlett allowed Bill Bean a free hand with his columns, and Bean proved to be a tough taskmaster who refused to let slips in grammar or fact get by him. Through the coming years, Scarlett willingly accepted and approved of Bean's thoroughness and did not mind his criticism. If Scarlett made a mistake, he apologized. "I congratulate you on your editorial eye," he wrote in May 1962, quite early in their relationship. "Sorry to cause you pain. Some of the slips are due to my newly acquired habit of composing directly on the infernal typewriter (a fearful way to serve the muse), some to ennui and others to plain ignorance (like Dr. Johnson). So keep after the 'fly specks.'"

Mastering his typewriter (an ancient machine of 1920 vintage) was a necessity both because of the volume of his writing and because he wanted to do his own typing. While his columns were still signed E.P. Scarlett M.B., Calgary Associate Clinic, his association with that body was purely honorary. His long-time secretary had offered to do his typing, but he felt he could not impose upon her.

At the end of the first year, he looked back, with some satisfaction, to "realize that I have done little more than publish a clutch of Quixotries,"[16] which is what he had intended to do. He had set out to lure readers to his column and had done so despite the clinical attractions of the journal.

In his original letter of acceptance to Bill Bean, Scarlett suggested that he might do some pieces in serial form. He started this in 1962 with a two-part column on words, but beginning in June 1963, he launched into a three-part column titled "The Head of Medusa" and subtitled "Contemporary Literature's Obsession with the Pathological." It is a brilliant analysis of the contemporary scene, especially in connection with the love affair with Freud that was affecting so many writers. In it Scarlett laid down his literary law in no uncertain terms, suggesting that such writing was not only an artistic but a moral evil. Few authors of the time would have enjoyed the analysis, and professors of modern literature, reading "The Head," would probably have dismissed it as a last gasping breath of Victorianism. It is unlikely that it would have been published in any journal devoted to literature. Its appearance in *Archives* is a credit to Bill Bean and his readers.

Broadly speaking I am suggesting that this preoccupation with the pathological, directly or by inference, constitutes an attack on nearly all that is decent in life and in society; that it plays up the brutal, the ugly, the greedy, and the selfish, and presents this as life; that it is, in effect, an attempt to exalt the baser side of human nature; and that, in biography, rather than burying the good of its subject with his bones, it proceeds to disinter the bones and magnify the evil . . . I am suggesting, further, that it is time that we called a halt to such practices as distortion of the truth. We should say plainly that these writers are setting up false standards; that in revolting against Victorian romanticism, they have romanticized evil; that they are befuddled exponents of the Freudian revolution; that they are destroying standards of decency for which civilization has fought since the beginning of time; that they are fostering a contempt for the dignity of man and creating a sense of lack of responsibility in the minds of their readers.[17]

His arguments were logical, although many readers no doubt disagreed with the premise. He attacked novelists such as Harold Robbins and literary deities such as W. Somerset Maugham and Tennessee Williams. It is not surprising that he would ask a medical journal to devote page after page to what he admitted might be seen as "a wearisome torrent of comment and obiter dicta on a host of matters dealing with life and literature." Physicians were supposed to be shock troops in a battle for a gentler, more humane society, and his words simply provided the vital "intelligence" all armies must have in order to win.

Not all of his columns were deadly serious or aimed at educating readers about esoteric bits of medical lore. In February 1964, the Zebulun Doctor provided an entertaining look at medical and literary hoaxes, including parts of a modern satire by a writer known as S.N. Gano of *The Leech*, a British medical student journal. Scarlett had been sent three copies, each from a different American city, and he enjoyed the somewhat sophomoric send-up of medical writing. He even reprinted some of the aphorisms of Hippocrates not generally found in medical journals, such as "Absence of respiration is a bad sign" and "Life is short and the art long; patients are inscrutable; their ignorance impenetrable, and their relatives impossible."[18]

In 1964 he published essays on Bach, Handel and Mozart which were good enough to give him something of a reputation as a musicologist in Europe.

In 1963 Scarlett wrote Bean a letter devoted, partly, to Sherlock Holmes, and he threatened to "break out into the Zebulun pages one of these days with a piece on my dear colleague, Dr. Watson, or with an account of my walks up and down Baker Street with note-book in hand." He had mentioned Holmes in passing, but in November 1964 he drew on his *Bulletin* piece to write about Dr. Joseph Bell, the figure behind Conan Doyle's master detective.

In February 1965, he wrote what should probably be required reading for any college student about to study English literature. The column begins by noting the difficulty a person of Scarlett's generation has in speaking to young people, including young doctors, because to them his experiences are time-shrouded and unknown. His real point, however, is that fewer and fewer people have any knowledge of the Bible. Without this knowledge, and that of the Anglican Book of Common Prayer, much of the colour of the English language is denied then.

Again, readers might have asked what all this had to do with medicine, but mostly they didn't. They realized that physicians had to protect their minds from needless narrowing. "The response to the Head of Medusa eruption has taken me aback," he wrote Bean. "I am buried for the moment under 'Missives.' But it is most encouraging – and I know that you share my satisfaction. Certainly there is still corn in Egypt."[19]

The letters from readers did not come only from doctors. A New York playwright wrote, not to criticize, but to cheer. One letter that brightened Scarlett's day was from Sir Francis Walshe, a British neurologist of note and one he especially admired:

> I received his letter with much the same feeling as if I had a communication from Mount Olympus – for he was one of the clinicians whom I adored when I was "walking the wards" of the London hospitals years ago. The relation of master & pupil has always remained with me, and I find it hard to resolve even if I am now over three-score years. I must pass on to you some sentences from his letter, not only because "praise from Sir Hubert is praise indeed . . ." but because they thoroughly reflect my own sentiments. "We have not a medical journal in this country that takes so high a view of its functions as does the *Archives of Internal Medicine* since Bean took over the chief editorship . . . No journal over here would publish your Head of Medusa, or even the kind of paper which in my old age it now interests me to write.[20]

The reaction to "Medusa" was not all good. At least one professor from the arts side of a university suggested to Bean that it was improper for a doctor to step out of his discipline and comment on literature as though he knew something about it. Bean sent on the letter and Scarlett responded with the comment that he had seen this sort of thing before and it had ceased to worry him.

Being editor to the Zebulun Doctor had given Bill Bean an idea: Why let these columns lie buried in medical libraries? What, he wrote Scarlett, would you say to having them published in book form? This suggestion appears to have surprised the author, although he was not displeased: "I confess to being a bit staggered – surely the stuff hardly merits that. If, however, you want to plead the cause before the Throne of Grace I shall not object. But don't worry about it. I am quite satisfied with the present dispensation."[21]

A little over a year later, Ed Jordan, editor of *Group Practice*, made the same suggestion. Scarlett decided he might go along, for fun, with the idea of a combined Jackdaw-Zebulun concoction and told Jordon to write Bean. Jordan did so, but Scarlett had second thoughts, worrying about throwing even more work onto the backs of his two editors. Perhaps, he suggested, the idea might be put to rest.

In fact, a third figure, Dr. Stanley Greenhill, professor of preventive medicine at the University of Alberta and editor of *Current Medical Bulletin*, a quarterly for which Scarlett also wrote, had talked about the possibility of a book. Given the Canadian record in such publishing, Scarlett figured this was unlikely to happen. He was right.

Despite Bean's best efforts, and both Bean's and Jordan's editorial work, no publisher could be found. As Bean wrote to one Texas physician who had independently suggested such a book,

> We have made five different efforts, with as many book publishers, to publish the Zebulun Column, or excerpts from it. There is a pitifully small market for this sort of thing and have got no takers yet and I am sure we will not; therefore you will have to get photocopies made if you feel like it. I hope some day to publish the correspondence I have had with Scarlett but this, of course, is looking far in the future.

A Scarlett "book" would have to wait until 1972. In that year, a selection of his writings, edited by Charles Roland, was published by McClelland and Stewart in Toronto.

Earle Scarlett's subjects were as wide-ranging as his library, but he usually wove medicine into the topic at hand. A classic piece was the seven-page treatise on "The Doctor's Wife" in the March 1965 issue. It might well have met with stern disapproval from the more militant feminists, although he tried to write on this subject with "all humility, without conscious prejudice, and with the absence of that implied male superiority which is so apt to infect a discussion of this sort."[22] He wittingly or unwittingly describes his own marriage when he talks of the doctor's wife's taking a back seat to medicine.

Circumstances and long-established custom of the days, before women charged the male barricades of society, of course made for the retiring wife. And in the case of the medical household the wife not only looked after the home, but brought up the children as well. The image of parenthood to the doctor's children was mostly motherhood, a tiny bit of fatherhood, and practically no parenthood as a joint enterprise. Patients came first. I recall discussing this matter with the wife of a physician of the older generation. She explained the hierarchy which was worked out in their household in the answering of calls. Children first, women next, then old men, next adult males (even if they were more likely than women to be overanxious about their earthly tabernacles), and near the end known cases of hysterical illnesses. Finally, she added wryly, the doctor's own family, who were left largely to the attentions of mother and God.[23]

His point in the essay is that doctors' wives are the unknown and unsung heroes of the medical battlefield. The husbands get the glory – although, as he wrote elsewhere, the glory was less than it once was – and the wives get left at home with the children. He delighted in the story of the self-important husband who said to his wife, "Dear, I wonder how many great men there are in this country?" And her reply, "Well, my dear, there is one less than you think." The essay's tone has something of heavy-handed chivalry, but it also has a great deal of love. The essay reflects a time when Scarlett, retired and working at home, had come to know and appreciate his own wife more than he had been able to as a practising physician. It was, if you like, a salute to Jean and the other women who had spent most of their lives answering the phone, raising children, apologizing for broken dinner engagements, and wondering why their husband was so late.

In September 1965, Scarlett received a letter from Bill Bean he did not like. The two friends-by-mail had forged an editorial bond the like of which Scarlett had not felt since Stanley's death. Bean's letter was disquieting.

First of all, there have been a few storm warnings appearing on the scene that I should alert you to, that you may make your plans accordingly. I don't anticipate any hurricane, but one never knows. John Talbott, editor of the *Archives*, has suggested to me in very direct terms that I have been contributing too many book reviews and editorials. This hits me just at a convenient time when I have run my backlog completely out and have to write a book on rare diseases in atonement for giving the Beaumont Lecture in Detroit this April. The book is taking up more than all my spare time ... Also, the notion that we are making this too much of a literary journal seems to offend those who enjoy reading medical writing, the dust of the desert but not the spring and oases. If I find I'm not permitted to let my views prevail, I'll certainly turn over the running of the journal to somebody who might or might not look with favor on Zebulun. I believe in any event, I shall aim to turn over the editorship of the journal to somebody else in about a year and thus I can make no commitments beyond that. The editorial content of the journal can be exerted in a very subtle way through budgetary considerations so there might be some retrenchment this coming calendar year and after. I hesitate to predict. Anyhow, I wanted to give you ample warning.[24]

Scarlett wrote back from Farnham, Quebec, where he was visiting his daughter Katherine and her family, expressing gratitude for the advance warning. Parts of Bean's letter, he said, seemed to suggest that his column could be causing problems.

I am concerned only about two aspects of the business. The first is that you should continue in editorial control. It would be little short of criminal if you bowed out. For, you have created something unique and precious. That is not my opinion alone, but I know that it is shared by a host of physicians who have found deep delight in what you describe as "the spring and oases." The second consideration is that the continued presence of "Zebulun," enlisted and serving under your banner, should not be an embarrassment to you. If getting rid of this incongruous little booth in the clinical marketplace of the *Archives* would secure your tenure in editorial office, I would urge that you give it its discharge papers ... I am serious about this matter. You have full permission to behead me any time.[25]

If Bean did go, he continued, he wasn't sure whether he would want to stay on. "I should feel very much like a forlorn unicorn in an alien garden."[26]

Dr. Scarlett in his office at the Associate Clinic.

Top: Scarlett in his greenhouse, 1962.
Above: Jean in the garden, 1968.

Earle and Jean with grandson Michael, 1969.

Earle and Jean relaxing on holiday, 1969.

Left to right: Hartley Scarlett (Earle's brother), Earle and Jean, in front of the greenhouse.

"The Kingdom of 409."

A family reunion, 1973.

Scarlett – with pipe, dogs and books.

Despite Earle Scarlett's pleas, Bill Bean was up to his ears in work and the *Archives* consumed too much of his time. He knew the guard was changing at his journal. It was a bad time to break bad news to his friend – Scarlett had just suffered a second serious coronary – but towards the end of August 1966, Bean wrote to say he had made up his mind.

> I have definitely come to the end of my editorship of the *Archives of Internal Medicine* and the new editor will be Carleton Chapman who has left Southwestern Medical School in Dallas to become dean at Dartmouth, a two-year school which is expanding into the ranks of a four-year school. John Talbott, the overlord of AMA journals, has asked me to stay on the Editorial Board, but I have not decided yet whether I shall do this. If I am sure I can do it without any unconscious overinfluence on Carleton Chapman, and he is a thoroughly and totally competent sort of person, I think perhaps I shall.[27]

In September, having emerged from hospital and able to tackle some of the work which had piled up on his desk, Scarlett wrote back that he understood.

> Your letter ... tolled a bell that moved me profoundly. Verily the end of one of the happiest chapters of my life! What I have been wondering is whether, having sailed for more than five years under your command, I should similarly "pack-up" when you strike your colors on the flag-ship? Or should I tail along, and wait for the diplomatic communique to transfer to some other fleet? To change the metaphor, now that the engine on the head-end is being changed to a new diesel-model, should I, as the caboose (of an antiquated Victorian design) not cut the coupling and be shunted to the rear where old models are allowed to rest in museum peace? I await your advice.

Scarlett's words were set in a slightly joking manner. Bill Bean had done his best to soften the blow, but the Doctor Out of Zebulun felt he was a bit like Turner's painting of the ancient sailing vessel being towed to the knackers yard by a steam tug. He was seventy years old, and while only a hospital bed could slow him down, the world and his sacred profession seemed to him to be losing their soul. Bean and he, together, were at least waging a rear-guard battle. For once he could not find the words to describe his feelings to his friend.

Bean, perhaps because he was younger, approached the Zebulun problem from a practical viewpoint. He decided to discuss the column with the editorial board of his journal when it met in November. Until then, he advised Scarlett, it might be advisable to hold off sending any more columns until he knew better what the future might hold. In any case, he still had some on hand.

The November meeting, however, didn't solve much. The whole thing, Bean wrote, was getting him down.

> I have no idea yet what fate is yours and what will become of Zebulun. I brought the matter up briefly at the last meeting of our Editorial Board in Chicago, but because Dr. Chapman will not take over active editorship of the journal for some time and because Dr. Talbott clearly wants a free hand in running it, and I think this is perhaps his prerogative, though I hope the home office doesn't really get control, things are very much up in the air. . . In a solemn and somber mood, I feel that what you and I have been able to do for the *Archives of Internal Medicine* in the last five years has established records of some kind, but beyond that, it puts before those with eyes to see the realization that medicine is more than money, the laws of mass motion and morbidity.[28]

Scarlett's apprehension that his services were no longer required was not lessened when he received a letter from John H. Talbott, M.D., containing his honorarium, $250. It confirmed that Bean had resigned and that Carl Chapman would take over in 1967. Since he didn't want to commit Chapman to anything, Talbott advised Scarlett that he might as well hold off for the moment.

> It may be that he will wish to resume your series at regular or irregular intervals, but this is for the future to decide.
>
> In the meantime, on behalf of the Board of Trustees of the American Medical Association and the Division of Scientific Publications, I am pleased to acknowledge your unique contribution to the *Archives* in recent years.

Scarlett's feelings seem to have been hurt by what he called a "curt, cold and most official letter." In a letter to Bean he wondered whether he might be reading too much into Talbot's words, but "I think he is rather pleased to be dropping one of Bill Bean's crew."[29] In fact, given that – as it turned out it – wasn't his swan song, he probably was ruled by his emotions. But he had to agree that Talbott was right – and Bean too, for that matter. Chapman must have a free hand and until that time Talbott and his colleague, Sam Vaisrub, M.D., were in charge.

By April 1967, however, Scarlett was able to write Bean (who probably already knew) that things had changed again.

> I must tell you that negotiations have begun to re-instate Zebulun. Dr. Vaisrub has written to ask if I would give my "approval in principle" to contributions in the old vein at less frequent intervals during the year, adding that the Department of Scientific Publications would "be in a responsive mood." Just what this diplomatic language means I do not know but talks are proceeding.[30]

By June he still wasn't sure what that "diplomatic language" meant. Chapman hadn't yet taken over and there had been no more letters. Scarlett decided to test the water and see if "approval in principle" meant a quarterly effort. He sent a column off in November, with a note to the effect that he was quite willing to scrub the column altogether. Bean offered to intercede, but Scarlett did not agree.

In March 1968, he received a letter officially reinstating him "in the Archives stable" on a quarterly basis. This seemed to have put him in a better mood, despite continuing heart trouble and his wife's serious illness.

> Although back in the Archives' stables, I have no notion when I will be making my first run under the new colours – looking back, I am amused to think that, following you, I was all prepared to be turned out to pasture. Indeed I had prepared a suitable farewell to Bogdonoff, and the new training stable, the text of which was, suitably enough, drawn from Deuteronomy 13:18: "Rejoice, Zebulun, on thy going out." But I had no chance to use it.[31]

He could not use this epitaph, and the response to his re-established column, notably to "The Profession in Contemporary Society," showed he still had his readers, including Bill Bean. In thanking him, Scarlett wrote:

> I am grateful for your always beneficent words concerning "The Profession in Contemporary Society" in the Zebulun corner. As always you are an old and alert watcher on the tower of professional integrity and freedom. The response from the clerisy at large has been heartening to say the least. There

are still, I am pursuaded, simple "righteous men" within the walls of the city of medicine to save it from destruction by the barbarians. Speaking of Zebulun, I am gradually getting acquainted with Bogdonoff, seemingly a most reticent man, but, I would think from his editorials, sound in his fundamentals. We are not in the same wavelength entirely, but he tolerates me, if quizzically. In this world of medical journalism I do miss our old Ed Jordan, but his successor, Mrs. Keenan, has been quite wonderful and tolerant.[32]

Scarlett wrote a few more quarterly "Doctor Out of Zebulun" columns, but without Bill Bean at the helm he and the magazine were not on the same course. He was, he told Bill Bean in 1971, not dropped as much as allowed to fade away.

Either way, a chapter in the history of medical writing had closed.

Chapter Eleven

In my profession all sorts of odd knowledge comes useful,
and this room of yours is a storehouse of it.
— "The Adventure of the Three Garridebs"

NINETEEN HUNDRED AND SIXTY-SIX PROMISED to be a vintage year for Earle
Scarlett. Foothills was about to be officially opened, and he was finally to
meet Bill Bean. For years he had tried to arrange a meeting and at last the opportunity had presented itself. Bean was to give the Sir William Osler Oration at the Canadian Medical Association meeting in Edmonton, which
Scarlett would attend. Gail Bean, Bill's wife, would also be there, and after
the meeting the two couples would spend a week at Lake Louise.

The arrangements, however, fell apart. On the twenty-fifth of May, while
planning for both the Foothills' opening and the Beans' trip, Scarlett suffered
a myocardial infarction and was rushed unconscious to the intensive care
ward of his old stomping ground, the Holy Cross Hospital.

For two days he lay "surrounded by the fearful instruments in the intensive treatment centre" of the hospital. As soon as he was able to switch to a
private room he began to write letters surreptitiously to his friends. Jean
wasn't especially pleased, but agreed to mail them.

The first, naturally, went to Bean, and the reply came by return mail. Bill
Bean would still drop down to Calgary, but to save his friend from too much
excitement he would come by himself and stay only a day. Scarlett was disappointed, but the prospect of even this brief visit from Bean buoyed his
spirits as nothing else could. "It was as if all the heavens had opened in one,
beneficial glimpse," he wrote. Moreover, he said, the separate letter Gail had
enclosed for Jean helped cheer her up during a very difficult time. His only

worry was that he would not get out of hospital in time for Bean's visit on the seventeenth of June. As it happened, he made it. And, as he had noted to Bean in a letter in June, he was beginning to wonder if his heart attack had changed his status from elder statesman to medical saint.

As fortune has decreed, I am having the strange experience of being canonized in absentia, much like a minor saint. I have already received in the past fortnight in the course of medical ecclesiastical ceremonies a life membership in the Canadian Arthritis & Rheumatism Society, and what is known as the Golden Jubilee Award of the University of Alberta for services to that institution. The Can. Medical Association award this next week completes a triple business. I am indeed in the odour of sanctity, and have made the grade without having to undergo the customary inquisition of the Devil's Advocate.

His doctors had reluctantly allowed him to have a typewriter by his bed. There wasn't much they could do about it; elder statesmen tend to get their own way. In this way, he managed a Zebulun column, "written in a cardiac posture with the re-enforcement of a pile of pillows."

Mrs. Coral MacDonald, a nurse who had taken classes from Scarlett, had also worked with him in the hospital. She now found herself looking after him and discovered that, like many doctors, he was not an ideal patient. He was typical of most coronary patients in one way, however. "He was like all patients, he couldn't really believe it had happened," she remembers. "Nor did he want to pursue the regime he ordered his own cardiac patients."

"The hardest thing about being ill," he told Mrs. MacDonald, "is to do what I have been telling patients all these years: you have to cut back all your activities."

Bean's visit took place as planned, in Scarlett's home, although Bean was able to spend only a few hours with his friend. Scarlett was still too weak to do more than talk from his chair, but it was a meeting neither man would ever forget.

It was the twenty-fourth of June before Scarlett found the strength to convey to Bean in a letter how notable the event had been: "I am still surrounded by an aura of elation, recounting to myself and periodically to Jean so many of the avenues which we explored in those blessed five hours."

Bean had given Scarlett a copy of his "oration," and Scarlett was delighted with his friend's words.

It is the best Osler oration yet. You are in top form as befits the subject. What a magnificent counter-blast to all the current heresies, cynical chatter, and weird slogans of the contemporary rag-merchants being shouted in the market-place! I like the way you plunge in and make straight for the mark – "is human life significant?", scorning the medical suspicion of metaphysics.

"No Freudian nonsense here," he concluded. "After all, it was faith, not ratty Freudianism that raised Lazarus."

It was no surprise to Coral MacDonald that Dr. Scarlett refused to take his own advice. He and Jean drove up to the mountains for a vacation and then he went back to work. He was still heavily involved with the University of Alberta and flew to Edmonton on 4 August for a full day of meetings. It was too much for him.

On 6 August he was hit by a posterior coronary. Three days later he regained consciousness to find himself again in intensive care and on oxygen. Ten days later, "by special Papal decree and nursing assistance," he sent a short note to Bill Bean.

Bean wrote back partly to congratulate Scarlett on surviving the second heart attack and partly because he had his own bad news to impart, that he was leaving the *Archives of Internal Medicine.*

It seemed to Scarlett that the fight against "the barbarians" was about to end, not in surrender but in retreat. Still, by September he was feeling much better, spoiled by visits from friends, fellow doctors and his old nursing staff – and snowed under by more work.

I have emerged. Navigating under strict orders, I am once again tasting the winds of heaven. It is true that I am partly hidden by a huge pile of administrative data "for attention," and a fearful array of reports & what not concerning University Commission matters. In looking at all this pretentious material, I have in triumph resurrected our old war-time slogan of Canadian Corps memory – Ça ne fait rien" (which in the early vernacular of those days we rendered as "Sin Fairy anne"!). That same far-off "adventure" of my generation has rendered for me also my motto for the next space of months: "Out of the trenches by Christmas!" . . . It only remains to note in the personal realm that I am coming along painfully slowly – but still "coming." Driven on a tight rein under Jean's eye, I get about a bit each day, am allowed an hour in the greenhouse daily, visits from one or two friends for an hour two or three times weekly, but no going out to such inquisitions as committee meetings or friends' homes. So I live in partial retreat – not in any penitential way – but in my home and private world where there is no confusion, exhaustion of spirit, despair, nor aloneness . . . Jean and I greatly appreciated your letter to her. She has borne a great burden of anxiety and executive work which has kept all these months bright and shining. For my part, now that I have, so to speak, two strikes against me, I shall play the game from now on with all due caution.

By October he was out and about again, although still under Jean's eagle eye. Since both he and his friend Smitty Gardner were rose afficionados, they would often chance upon each other in the rose garden of a nearby public park. When Jean thought they had had enough, she would firmly steer Earle back to his house and his study.

Towards the end of October, as Jean relaxed from the strain of looking after her husband, they decided to drive west. Heavy snow held them up at Banff, so they enjoyed a lazy time in the town where they had taken their first trip together so many years ago. There he wrote to Bill Beart, urging him not to work too hard and noting how, after his own semi-retirement, he had been able to enlarge his horizons and become re-acquainted with his wife.

Zebulun worried him, since he still was unsure of its fate, but he at least felt that he and Bean had made a contribution to the philosophy of the practice of medicine.

In April 1967, Scarlett received a letter from Bean which brought back old memories. In the 1940s one of the journalists who dropped into Scarlett's study was Richard J. Needham, then an editorial writer for the *Calgary Herald*. Needham was cut from the same cloth as Bob Edwards, the iconoclastic editor of the *Calgary Eye Opener*, which is perhaps why Scarlett insisted on calling him "Bob." Needham was a brilliant writer and essayist and was eventually hired away by the *Globe and Mail*, which he usually called, in print, "the mope and wail." Bean had come upon a Needham clipping on a favourite topic of the columnist, love and women, and had sent it to Scarlett as an item of interest. Scarlett replied that he had indeed found the column interesting, but that it had saddened him slightly.

The article is vintage Needham. I could talk to you for an hour or more about this lad – a genius manqué; an Englishman born in Gibraltar who came to Canada in his 20s (a Blue School Man), cut his newspaper teeth in the West; and later went to Toronto where he had worked out his strange destiny, and now (he is in his 50s) is nationally known. The irony of the article in question is more appreciated by Jean and myself than by the plebs. Bob always went it alone (it was years before I knew he was married and had two little girls). In the years he lived here he was never once seen in public with his wife. Now that the children have left the nest the strange, diabolical, Villonesque strain took over. He left his wife and home and lives in one bare room in the bohemian, "beat" quarter of Toronto known as Yorkville. As a result, we are no longer in the same orbit, but I am hoping that with the charity of advancing years a reconciliation may come about. In the meantime he remains the most *individual* (in every sense of the word) that I have known. In the course of midnight palavers (he always did his writing after 1 a.m.) he used to say that three persons constituted an intolerable crowd. Now we read his homily on love and women with the above in mind.

In a further letter, in June, he added:

The strange parallel between my good friend, the Voltairian Richard Needham, and his mentor Albert J. Nock, has haunted me. I can only explain Needham's present conduct as a protest against our North American way of life which he loathes and derides. Yet I know all the while a sense of decency and a faith in the sovereign value of love are at the helm, no matter how cunningly disguised.

Scarlett had always enjoyed philosophizing about life. Now, driven perhaps by his own precarious physical condition, he began to look at himself. He wrote to Bill Bean in June 1967:

For many years I have suffered from a curious disability which plagues me three or four times a week (doctors always have ailments that defy the conventional diagnostic categories). Almost immediately after breakfast to which I have sat down with "shining morning face" and in good spirits I am seized by a violent "hay fever" sort of thing – eyes and breathing passages in an extreme state of deliquescence – which lasts some 20 minutes and which completely incapacitates me . . . it passes as suddenly as it comes on. All experimental and theoretical attempts to pinpoint the cause have failed. Long ago we wrote the business off as an allergy to the prospect of work in the day ahead (a syndrome which has never been described in the voluminous allergiacal literature). In recent weeks I have had a revelation in this regard

– and am now driven to believe that the phenomenon is a reaction on the part of my autonomic nervous system to the ghastly apocalyptic state of the world which somehow my organism must face for another 24 hours. It is possible that this strange reaction may be heightened by the fact that I am an old constitutional owl (at midnight I can read Ecclesiastes without flinching) but who cannot face the lark's glad morning hours without dissolving into complete impotence.

Thankfully, he added, he had Jean to bring him back to normal. "Jean, thank God! is an indomitable lark who provides an unfailing and admirably anti-congestant effect. As I often tell her, it does not take a woman six days to create a shining world for any man."

The year 1967 was Canada's centennial, and Katherine kept Scarlett up to date with Montreal's Expo'67, which Scarlett saw as a "Canadian renaissance." He was happy to see Quebec as a site for this great exposition, since his respect for French Canadians dated back to his war years. He liked to remind people that he had fought alongside "the legendary Vingt-Deux brigade." While Katherine's letters from Montreal were happy, the letters sent east were not always so. On the fourth of September he was back in the Holy Cross with another coronary, a small anterior infarction. It was not serious, but in late October he suffered yet another. He wrote Bill Bean two weeks later:

> Two weeks ago lightning struck again without warning and I was lugged off to my old hospital with what proved to be a *small* infarction. I am now, due to a most generous change of heart on the part of my colleagues, home again, sitting at my "piled high" desk in slippers and a glamorous new housecoat which Jean made for me, fortified by a new Dunhill pipe, brisk autumn air without, a good fire in the grate, and a burgeoning sense of well-being. Once more, under the anxious eyes of my hard-driven home nurse Jean, I am plodding along the carefully-marked road of convalescence.

In December, Scarlett found himself honoured again when he received a formal letter from the Calgary Public School Board stating it was the board's unanimous wish that a new high school be named after him in recognition of his "dedication and service to your chosen profession and outstanding contribution to the educational and cultural life of Calgary and Alberta."

Scarlett replied in typical form: "The name proposed – the 'Dr. E.P. Scarlett High School – is quite agreeable to me, though I must admit that it scans wretchedly, and does not lend itself to any poetic outbursts."

During the same month, he decided to combine health and pleasure by travelling to Montreal to visit his family and get checked over at the Montreal General Hospital. He was feeling fairly well, but wondered about possible therapy. His Montreal colleagues put him through a thorough medical examination and were "guardedly optimistic" about his condition, provided he took retirement seriously and shucked his university and other responsibilities. He was able to spend half a day at the Osler Library.

As he was trying, unsuccessfully, to figure out what would happen to the Zebulun column, he was discomfited to learn that *Group Practice* was also going through a change of captains.

It is an interesting coincidence that Ed Jordan . . . has retired as director of the Am. Assoc. of Medical Clinics and editor of *Group Practice*. The office has been moved to Alexandria, Va., and the new director is someone called Dr. Edward M. Wurzel with whom I have no acquaintance. I have heard nothing concerning what will happen to *Group Practice* and "The Medical Jackdaw." So closes the happiest of associations with a great gentleman, Ed Jordan.[1]

Although *Archives* was the better-known journal, he had stayed loyal to *Group Practice*; thus, his humour was improved when it appeared the Jackdaw would be able to survive its tree being shaken. Dr. Scarlett was, however, about to receive his most severe shock to date. On the thirtieth of January, while Earle and his wife were out shopping, Jean suddenly collapsed. While he had been able to suffer through his own heart problems with reasonable equanimity, this was harder to bear. Because of his weakened heart, Scarlett's own physicians wisely insisted he take up residence in the Holy Cross, where his wife had been admitted. Writing to Bill Bean "in heavy distress" from the hospital the next day, he described what happened.

When we got her here into hospital it was found she has suffered a massive anterior coronary infarction, one of the worst I have ever seen. She is still teetering along, but her condition is grave. Trouble in sustaining her general circulation . . . Everyone has rallied around. My clinical cardiological team have been wonderful. After spending some time with me last night at home, they moved me back in here this morning. I am on the floor above the intensive care unit where Jean is, and can slowly go down to see her and check on the observations once in a while. The boys felt that I should be in here as a protective measure, as this business and the sub-zero weather are a heavy strain on me. We can only hope that the good fortune that it has been the happy lot of Jean and I to enjoy for more than 45 years will continue to hold.

Earle was allowed to go home after ten days. Katherine and his two grandchildren arrived to take charge of the house. He was, however, kept on a strict regimen and ordered to rest. Jean slowly made progress and returned home in early March. With the help of their housekeeper – his daughter had rejoined her husband who was studying in Yugoslavia on a Canada Council grant – she returned to health faster than her husband after his first attack. By April the two were able to go outside for walks.

His condition appeared to have stabilized, but he was under strict orders to take life easy. He continued to write his columns and to "smoke his pipe to the glory of God." By the summer Jean could again drive the car and life began to return to normal. They were able to visit the mountains, and by September they rejoined their recorder group to play the Bach and Elizabethan madrigals they loved.

He was no longer an official "cardiac patient," but he was still on anticoagulants so he had to avoid anything that might cause bleeding. He was also taking part in a long-term study of the use of the beta-blocking agent propanolol (marketed as Inderal) which, he told Bean in September, was soon "to be released under some poetic trade name."

In May 1969, he was back in hospital, but this time it was not for his heart. For some time he had been troubled by an inguinal hernia and had asked that surgery be done to correct it. His doctors, with some reluctance, sanctioned the operation. Although he came through without difficulty, recovery was complicated by a wound infection, and much to his annoyance, he was kept in the Holy Cross for about three weeks.

Construction had begun on the E.P. Scarlett school and his home was becoming something of an inadvertent headquarters for the project. Scarlett wanted to be in good health to contend with a certain amount of confusion. In one instance, a large truckload of radiators for the school's heating system was delivered to his home. The driver's explanation was reasonable: "409" was the only address for an E.P. Scarlett listed in the phone directory.

In the spring of 1970, Earle Scarlett was overjoyed to receive a horticultural gift. On two occasions he had planted seeds from the tree on the Island of Cos under which, according to island tradition, Hippocrates had taught. None of the seeds had germinated. His friend, Dr. Bill Gibson, chairman of the department of medical history at the University of British Columbia, had brought Scarlett a cane from the tree. The cutting, safe in the Scarlett greenhouse, had begun to produce leaves. The plant, representing Greek culture he loved and the history he treasured, was one of his greatest prizes. In return, Scarlett sent a cheque to the International Hippocratic Foundation towards the construction of a building on Cos.

He was still involved with the Foothills Hospital and was depressed over the increasingly shaky status of "The Medical Jackdaw" when he wrote to Bill Bean that spring to congratulate him on his appointment as the Sir William Osler professor at Iowa University.

You will have noted the sad decline of *Group Practice* – always a feeble affair but now sadly caught in the turmoil of periodical and newspaper publication. In the confusion the "Jackdaw" has not been evicted from the nest, but for a time being locked in his cage. It is a curious business. I have received no communication from anyone calling himself an editor or managing editor, but have had letters from someone called "Advertising Co-ordinator" advising me that I will be called when wanted. Seemingly I am relegated to the pharmaceutical products category. I suspect I am to be a victim of the new method of dealing with contributors – just forgetting them – a practice that Bogdonoff followed in dealing with the "Zebulun doctor." Being an old soldier, and thus slated not to die but simply fade away, I should fall fairly well into this modern dispensation. *Cela ne fait rien.* I maintain a tangential acquaintance with medicine. We are launching a medical school in connection with the University of Calgary, the first class to come in this fall. The Foothills Hospital, of which I am a member of the board, will be the clinical core, so we are working with the new faculty to get things away in proper style. At the moment we are engaged in close combat with the residents and interns who, in the matter of salary, have developed delusions of grandeur ... Jean and I continue to go along at a modest pace. On the whole we are pleased with the record of the past year. The dogs are a great joy; our reading aloud continues to be a joy; our music flourishes; we have enough work for health and sufficient leisure for the appreciation of beauty and music.[2]

The year 1971 began with Scarlett in a mixed frame of mind. He had many reasons to rejoice, including the vigour of the pride and joy of his greenhouse, his cutting from the tree of Hippocrates on the Island of Cos. It had grown chin high and he enjoyed fighting the North American pests which were doing their best to kill it off. On the negative side, his career as a writer seemed to be winding down, at least as far as his "Medical Jackdaw" column was concerned.

My own bit of extra-mural writing has dwindled to nearly nothing. *Group Practice* is still in difficulties and while I am retained on the reserve, no firm arrangements have been worked out. Undoubtedly, as is the case with so many professional and quality periodicals, the Muse *Advertisa* has taken over. Just recently the Association of Group Clinics informed me that the editorial function now rests with a "public relations firm" of Selvage, Lee and Howard of Cleveland (and the association HQ is in Alexandria, Virginia!). So I find an increasing inclination not to be dragged at the tail of the *Advertisa* horse, and will probably get out of the circus ring and will write "30."[3]

His heart problems, with continued use of propranolol, seemed to be abating. He was walking further and going out more. His school had been completed, and had opened for business on 8 September 1970. Although it was not yet dedicated, Dr. Scarlett was a frequent visitor. He was to be considered, he told students, "not an ancient drop-out but a young drop-in." All the while he and Jean were working hard at the recorder. As he told Bean:

A while ago we played with our Recorder Consort of 12 members in a public concert at the university, teeming with "The Renaissance Singers." With our white hair we provided a suitable renaissance touch. Then in mid-December I made my debut at the new senior High School which bears my name. I was bewildered by the size of the place and was delighted to find it has the largest library of any school in Calgary. I toured the place, met the staff at tea, spoke to a crowded assembly hall, and topped it all off by standing for two hours or more at a reception talking with the youngsters and answering questions. The result – my faith in our young has been given a big boost. But how is one to keep in touch with this younger generation and the current goings-on? Certainly not through the medium of contemporary "literature" (I apologise for using the word in this context) which drives one to despair.

Jean and I in self-defense have turned to a device – we keep on our study-desk a half-dozen "proven" books – At the moment they include the essays of Montaigne, Charlotte Bronte's *Villette*, and the *New English Bible*, which we read aloud – munitions for fortitude. We have done the same thing with music.[4]

On the twentieth of June, Earle and Jean Scarlett were driven to the Canyon Meadows district of southwest Calgary to open the E.P. Scarlett High School. His dedication was short: "I dedicate this school. May it always be a kindly nurse, a place of humane learning and the home of the high-hearted and the highminded."

Describing the events to Bean, he noted:

The official opening of the Senior High School which bears my name extended over a week or so, with the usual pomp and circumstance. I was particularly pleased that my profession presented the school with a large framed portrait of myself. Nothing gives a man more satisfaction than to have thus expressed the regard of his peers. The students of the school presented me with a set of silver goblets from which I may (and have) poured suitable libations to the gods. For my part I presented the School Library (a magnificent area) with a mural setting out the motto which Erasmus used in his library – all in Erasmian Latin, which has caused immense curiosity among the students and will, I hope, stimulate interest in the classics.[5]

The quotation, *Assiduus sis in bibliotheca, quae tibi Paradisi loco est*, was a Scarlett favourite: "May you be busy in your library which for you is paradise." He provided the librarian with a translation for the benefit of those who did not know the language. According to the assistant principal, Tom Sorenson, Scarlett always hoped Latin would be taught at the school, but it never has been on the curriculum.

Scarlett later congratulated the school for calling its yearbook *Compendium I*. "You have done well to draw on the perennial Latin for your title," he wrote in his contribution to the book.

Compendium – literally "that which is weighed together, an epitome or, as the Romans would have said, *"multum in parvo"* – has an entirely appropriate ring about it. (Strange, isn't it, that Latin which has been pushed out of the school's front door slips round and comes in the back!) In these pages

you will be listing the names and achievements of your school company.
Might I request that I be listed in this register as an Ancient Mariner who
has, I hope, more wrinkles in his face than in his mind?"

Part of the reason the opening lasted a week was that the school threw a
seventy-fifth birthday party for Scarlett. It was a good birthday with family
and friends, but it also bent his thoughts to the passing of time. The idea of
turning seventy-five filled him with "a great sense of wonder"; he wrote
Bean that, "thinking about it all, I am more than all else aware of the terrifying acceleration of time which aging brings, a sensation felt on the pulses
which is intensified by the incredibly rapid tempo in the world about us."
Earle Scarlett deserved the honour of having a school named after him.
The title "educationist" bestowed on him by professionals in the discipline
was merited – and he contributed words to each yearbook for the decade to
follow. He was not, however, a happy educationist. The course the secondary school system was taking worried him deeply. A "revolution" in education was well under way by 1971, and part of it was directed against the classical disciplines he felt were so important. Because of this, to him the school
was a source of both pride and worry. By bearing his name; it was a memorial to the impact he had made on the community he called home. But, when
visiting it from time to time, he did not, as he had hoped, find students and
teachers engrossed in the joys of good literature. Although his visits to the
school grew increasingly rare as his physical condition deteriorated, he often
talked of them and of the resulting feelings of despair and delight – delight
at detecting the occasional spark of enthusiasm for history and literature
among the students and despair at the road down which education seemed
to be taking them. His yearbook contributions continued to lay down his
own philosophy in a way he hoped might move "his" students.
"He was really concerned about the lack of pure academic pursuits,"
Sorenson notes, adding that Scarlett thought schooling was being turned into
a job-training exercise and that students were narrowed because of it.
One student, who commented that what he was studying might not
have a practical application, was told that education was useful for its own
sake. "Education is what remains after you have forgotten all the facts you
ever learned," Scarlett replied.
Several times a year during the 1970s, Scarlett would take a cab up to the
school and have tea with some of the teachers and spend time in the library
talking to students. Lunch would be brought into the library, and student

leaders came to eat with the patriarch. On one occasion, Sorenson noted, a student asked how old he was.

"I'm an octogenarian," he replied.

After a short pause the student tentatively asked what "octogenarian" meant. The ensuing lecture and the need to know the Latin roots of English lasted about fifteen minutes. His last appearance at the school was for its tenth birthday party in 1980. He left soon after he arrived, driven out by loud rock music. Scarlett made his last contribution to the yearbook in 1981, and in a way it summed up everything he had written for the school in years past.

> There is in the heart of each of you a little corner that contains and nourishes most of the precious things of life – imagination, sense of beauty, memory, honour, gratitude, compassion, reverence, and so many other precious tiny things. This is also the realm in which the peaks of your inner life are situated. It is "the piece of Divinity in each of us, something that was before the Elements, and it owes no homage to the sun." Tend that corner faithfully. Visit it daily. Defend it fiercely.

In the following yearbook it was noted that Dr. Scarlett was seriously ill in hospital. The editor wrote, "We owe this man a great amount of caring and respect. Next time you pass by Dr. Scarlett's portrait in the library, pause and give him a moment of your thoughts . . . he'd like that."

He would indeed. He would, in fact, enjoy the knowledge that the student was in the library.

In the summer of 1971, Scarlett reflected on his school and on his career as a writer. In his August letter to Bill Bean, he noted:

> The Medical Jackdaw, after appearing lately in the most atrocious format, is, as you predicted, retiring into periodical limbo along with the Journal. They are, they tell me, planning a phoenix act – the new creature to rise from the ashes about January, 1972, under the professional incantations of commissioned experts in the art of the tenth muse, *Avertisa*. I doubt if I shall rise again on that day. Jackdaws have only one life.

Much of 1971 had been a good year for the Scarletts, but, in his own words, on 22 November "the furies once more overtook us." A few hours after arriving home from a trip east Jean suffered a cardiac infarction and collapsed. In his Christmas card to Bill Bean, a very worried Earle wrote that "the infarction this time was smaller than before but involved the sinus node, so that we had to battle a bewildering succession of aberrant rhythms. After eleven days in Intensive Care, she was moved to a quiet room and then, in the spirit of the season, was transferred home a week ago."

For Earle Scarlett, 1971 had begun well and ended badly. The final "Jackdaw" had appeared in May, which brought his total of printed essays and other writings to 451. His wife was able to return home for Christmas but her condition continued to worry him even though she steadily improved to the point where she was no longer housebound. One thing he could look forward to was the publication of his first – and only – hard-cover book. Both Bean and Jordan had attempted to interest U.S. publishers in taking on a collection of Scarlett's writing, but without success. Charles Roland had had better luck in Canada. With Scarlett's help, he had sifted through Scarlett's vast collection of writings to find items which would give readers a true impression of that scrappy and erudite medical philosopher from Zebulun. A good-sized collection emerged under the title *In Sickness and in Health.*

Numerous demands on Earle's time took him away from his pipe and his books. Especially in the spring, everyone, including the occasional member of the press, wanted his opinion, and while Scarlett was always ready to oblige, he often felt tired.

Since about June 1 circumstances and events have stretched my limited physiological capacity to the ultimate so that I have been obliged to adopt every strategical means possible to keep going. In this I have succeeded and am presently in fairly good order. The dog days, visibly demonstrated in the habits of my two Scotties, have not assisted in giving one the necessary lift but merely contributed to a certain fuzziness of soul. There have been a multitude of interviews, old patients and friends coming in to discuss their troubles, medical school conferences, sessions with people from both the Provincial Government Archives and the University of Alberta Archives Division (as an "old-timer" I seem to have come to be regarded as an oracle!), and press interviews (reporters have discovered that I am still alive).[6]

In Sickness and in Health was published in 1972. Charles Roland, whom Scarlett described years later as a "fellow worthy of medicine" – as high an accolade as he could pronounce – had done his work well. It contained twenty-three essays drawn from *The Canadian Medical Association Journal,*

Archives of Internal Medicine and *Group Practice*, some profound, some funny, all worth reading.

In his foreword, Roland explained the *raison d'être* of the book. A crucial element in society, and one which was out of fashion, was "culture," which he defined as "that state of understanding and empathy with the works of man, scientific and artistic (as if these adjectives were polar!) which permits one to appreciate and learn from the past while still seeking to improve our knowledge and understanding for the future."[7]

Roland's words serve as one of the best working definitions of something so diffuse as to be almost indefinable. They also explain why he felt impelled to edit the book. He saw Scarlett as a healthy dose of preventive medicine for the endangered "culture." If reading this book does not "make you cultured," he wrote, "it will make you know what a cultured person is: Earle Parkhill Scarlett, for example."[8]

H. Earnest MacDermot, M.D., former editor of the *Canadian Medical Association Journal*, caught this idea in his preface when he wrote that "variety in life is what this collection is all about." It reflected "an unusual mind which, to use one of his own phrases, 'is rich in the juices of life.'"[9]

Fittingly, the book included Bill Bean's "appreciation," a thumbnail sketch not only of Scarlett's life but of how his thought had affected Bean and had convinced him to bring "Zebulun" to the *Archives*.

Scarlett himself contributed a short piece on the message he wanted to get across in his own writings. Partly, he wrote, his essays urge his colleagues not to forget that while a doctor moves in the realm of science he or she must also find time for questions science has ceased to tackle: "the fundamental problems of the meaning and purpose of life."[10] The medical person, he noted, has a difficult task. "He is trying to keep his balance as a human being in a civilization which conducts its worship in automobile showrooms, does its singing in commercials, lives on catchwords, gathers various impedimenta about him, and calls it 'gracious living.'"[11] But those who question a "culture which is mostly intellectual straphanging" will find rewards greater than they can imagine.

How many readers were converted by the book is impossible to tell. Certainly many of those who read it had already been influenced by the pen of the writer from Zebulun. Its publication came at a good time for Bill Bean, who could recommend it to those who wrote him complaining that the column had disappeared. A California doctor, Mayo W. Smith, wrote that he had subscribed to the *Archives* just so that he could read Scarlett's column.

"Such a prolific mind and pen makes a beautiful partnership and should not be stayed," he wrote, thanking Bean for information about Roland's book.

Scarlett proudly sent a copy of *In Sickness and in Health* to his school library, inscribing it with these words:

> This book is in part secretly dedicated to the Library of the Dr. E.P. Scarlett
> High School which quite properly stands on Canterbury Road – the road of
> the pilgrims . . . I should like to think that from the vantage point of a little
> corner on these library shelves, and from the fastness of these pages, I may
> look out on the life passing through these precincts. This will provide a
> strong cordial for longevity and a viaticum for the long future beyond.

Earle Scarlett was already in many medical libraries, enclosed in journal pages, but now he could stand in libraries on his own. The book was in a real sense Scarlett's final bow. His only published writings from this time were to be those in the E.P. Scarlett High School yearbook.

Mentally he was as active as ever. Physically he had held "the furies" at bay as best he could, but he had to admit his damaged heart no longer allowed him to behave as he wished.

Chapter Twelve

You don't mind the smell of strong tobacco, I hope . . .
— *A Study in Scarlet*

LIFE FOR THE SCARLETTS WAS now one of guarding their energy, of husbanding their strength. Earle was still in demand as an elder statesman and as a speaker, but all outside work was becoming increasingly difficult. He usually felt too tired to go out, but he had his pipes, his John Cotton tobacco and his study. He also had his garden and his greenhouse, and as physician and classicist, he was excited that his "Hippocratic tree" from Cos was healthy and growing rapidly. By 1973 it had reached six feet and was weathering all the pests the prairie could throw at it.

Never conventionally religious, he was growing more and more worried in the 1970s about the increasing lack of religion. He had become more aware of what he called the paradox of time that, while the "house in which the spirit is a tenant" is ageing, the "identity" of the occupier does not age. As far as he was concerned, he was still the young man who had set out for that great adventure in the Detroit hospital.

There were no more hikes in the mountains. Slow walks about the neighbourhood were as much as he could take. Travel was mainly by train; as Sherlock Holmes had advised in "The Adventure of Black Peter," "Try the Canadian Pacific Railway." He was delighted when, on one eastern trek, he ended up in a railway coach named for Sir William Osler, complete with a plaque at one end explaining who Osler was.

The "house" was in perilous condition. In May 1973, he suffered another heart attack and again ended up unconsious in the Holy Cross intensive care unit. His doctors decided to discontinue his medication. He soon began to suffer episodes of severe pain in the chest which he termed *status anginosus*. Earle and his physicians decided the pain could be due to his being cut off from the propranolol after having used it for more than six years. When the drug was restarted the pain disappeared. Scarlett was amused when warnings against sudden removal of the drug began to appear and wondered whether that was his contribution.

This year was the golden anniversary both of his marriage and his graduation. He was now permanent president of his class and had hoped to attend a reunion in Toronto, but his doctors put their collective feet down and forbade the trip.

Life was still good, however, for both Jean and Earle. Their greatest pleasure remained reading. They had returned to the classics, especially Dickens, for they had little to say in favour of contemporary literature. There were, Scarlett insisted, no great writers left.

He wasn't writing as many letters as he once did because his energy had to be hoarded, but correspondents continued to write to him. Mrs. Margaret Duthie, the librarian at his old clinic, helped him out with the requests for copies of this or that Zebulun column. He also received the occasional letter from young physicians, which bucked him up. It seemed that the humanitarian spirit of medicine was far from dead. One of these letters he shared with Bill Bean:

> Every week brings some excitement. Imagine – getting paid for leading a life so full of interest, challenge, hilarity, heartbreak, and, all the while, working with people on the team who share one's goals. We have our share of the callous, reckless, feckless, haughty and earthbound, but less so, I am sure, than many other groups.

"There is still balm in Gilead, my lad," Scarlett commented triumphantly.

The winter of 1974 strained the health of both Earle and Jean, so they spent March and April in Victoria, B.C. Jean's health improved but Earle's hope that the stay on the island would help him regain his strength did not work out. On his return, he cut out virtually everything in the way of speeches or public appearances and wrote nothing other than the occasional letter. The Scarletts allowed their lives to revolve around their house. They did not, though, live as hermits. They kept up with their reading and their recorder playing. Earle encouraged both students and teachers from his school to drop into his study for a talk. He was pleased when medical students came to him to discuss their problems, for he enjoyed young people and an audience, especially one that would listen to his lectures on the role of medicine.

Jean's weak heart continued to alarm him, although, of the two, Jean seemed the more healthy. Visitors to their home found her as dynamic as ever. But in January 1975 her condition suddenly became worse. She fell into a coma, and on the twenty-third of the month, she died.

To Earle it was as if his own life had ended. His physicians put him in the hospital, where he spent a fortnight in a state of semi-shock. His family rallied around and sustained him, but sooner or later he had to return to 409. He had a good housekeeper but he was now alone. The only thing which kept him going was an intense feeling that his Jeanie was still with him in 409.

In July he travelled east to Cobourg, Ontario, where Jean's ashes were placed in the family plot. He arranged for the ashes of Jean's sister, who had died five years before, to be moved to the same plot.

After the service he and his brother, who had retired and moved from Toronto to Fergus, Ontario, drove to many of the scenes they remembered from their youth. He then went to Saskatchewan, where his daughter Elizabeth and her husband, Daniel, had bought a farm near Saskatoon. Elizabeth had always been happiest among animals, and he was heartened to watch her in her new surroundings.

The one event which softened the blow of his wife's death was the birth of a daughter to Katherine. The baby was named after Jean.

His trip left him with a mass of impressions and memories his mind was able to work on and sort through. At the same time, he found it difficult to come to terms with Jean's death. The void it created grew rather than diminished as time passed, but he was able to write Bill Bean in October and reassure him that he was bearing up to his loss.

You, my dear friend, must not think that the world is too much with me, that at a time when the nations rage furiously together, and the daily paper is like a Newgate Calendar, I have bowed to the popular heresy of thinking that these are the fundamental things in life. Not at all. That way lies despair. The fundamental things are elsewhere and secure in the heavens and the hearts of men. And the greatest of these things, quite simply, is love. Love, as I now proved in the pulses, is a thing not affected by considerations of this world, which are "the rags of time." Love, sublime, unique, invincible – leads us to the infinite and the eternal. (I really believe that Jeanie wrote this last sentence – it must have come from one of her letters.)

Bean wrote back such comforting words as he could. He also pleased Scarlett by noting he had gathered their correspondence and all of Scarlett's writing he could find and had bound them so that he could read and reread them.

Scarlett's spirits lifted only slightly. Nineteen seventy-six began badly. To him, "time is the antagonist and so far it has been a despotic victor." His health had deteriorated to the point where his son, Robert, came to Calgary to discuss both Scarlett's future and his estate. His greatest fear was that he would get too ill to live without constant medical attention and would have to leave 409. His housekeeper, because of illness and family problems, was forced to leave early in the year. This meant Scarlett was alone and he had to admit that "the business of overseeing the household, rattling about in this big house alone (except for my faithful Scottie dogs), doing shopping, keeping up my various obligations has just been a bit too much."

A number of things kept him from sinking completely into despair. One was the feeling that Jeanie was still a presence in his life. It was not that he saw her as some sort of ghostly figure but that she was still alive as a part of his being. He created what he called a "shrine" to her memory on his study mantle, pictures of her surrounded by fresh flowers. This helped him come to terms with his grief.

His friends also came to the rescue. Besides running errands, they would drop in to spend an afternoon or evening in talk, adopting a rotation system so that Earle would have frequent company. He gradually reintroduced himself to his recorder group, which came by every two weeks for an evening of baroque music.

"I could not live without music, poetry and my library," he would say. With those three and his friends, especially Smitty Gardner, he could survive.

As Gardner wrote in the *Canadian Bulletin of Medical History*, "Although he was devastated and so alone with the sudden loss of his 'dear Jeanie' in 1975, a visit to his study was inspirational, his conversation stimulating, gentle and punctuated with classical, musical and medical references."

By March, Scarlett was still depressed. The house was clearly too much for him and his Kingdom of 409 looked as though it would indeed fall. Instead, what he called "a miracle" happened. In April one of his nurses told an old patient of his about his condition. Mrs. Willans, a widow, arrived on his doorstep and offered to be his housekeeper on the condition that she could consider 409 her home. She took over the running of the house. It was as though the universe had changed and he had received a new lease on life.

He began to take an interest in his school once more. He spent June and July with his daughter and granddaughter in Montreal. He also visited his sister in Winnipeg, where he arranged a reunion with six of the nine surviving members of his Manitoba graduating class. He even went south to visit Robert in California. The two of them indulged in "an orgy of music," but only after Scarlett had insured himself against what he half-jokingly called "the terrors" of American medical costs.

His health, everything considered, wasn't bad, although he liked to brag he held the world record for coronary "incidents." But he did not have enough strength even to write to Bill Bean, who was now director of the Institute for the Medical Humanities at the University of Texas, a posting Scarlett enthusiastically supported.

In July 1977, after a six-month silence, he finally managed to get a letter off to Bean. He said, "As a baptized man constantly trying to be worthy of a tiny measure of divine grace, I am horrified by the long hiatus in our legendary dialogue. I have encountered an 'energy crisis' in myself – physical insufficiency, and my course has been whimsical – up and down. Worst of all, writing has become a burden." Worse, he wrote, he was having trouble keeping his mind on his reading. He was not, though, ready to say his *nunc dimittis*. Instead, he had arranged a new régime.

I have contracted my periphery and within it carry on with the old essentials at a slower tempo. The result is that I feel more secure and a good deal better. In this achievement I have been nobly assisted by my housekeeper, Mrs. Willans, who is peerless. She has made this abode a home again and is a constant source of good cheer and inspiration. Without her I should have sunk ... I have a task-force of girls at the Public Library who every fortnight appear with a bundle of books. I dip into my own library at stated intervals to renew my faith. My music flourishes – both recorded and recorder-playing with musical comrades – thereby gaining "the evidence of things not seen."

And my dogs . . . are in good fettle and as companions set me a good example.

Even though his own writing was sparse, he appreciated hearing of Bean's crusade for the basic values and ideals of medicine. This, "in quiet corners," is also going on in Canada, he told Bean, and that gave him reason for hope:

> They are providing a ringing affirmation that is needed – a full-throated denial of the doctrines that the ignorant, head-shrunken idiots of our time, psychologists, sociologists and two penny philosophers are preaching in the marketplace, that there are no eternal values and that the good life is a lost dream. May St. Anthony's fire scorch their snouts!

That summer Scarlett stayed home and let the world come to him. His son, Robert, dropped by on his way to attend the World Harpsichord Festival in Bruges (Scarlett always told visitors that his son, a solid-state physicist, had an avocation as a professional harpsichordist). Katherine and his granddaughter arrived, as did a delegation from his Saskatchewan family.

In this way he was able to rebuild his life around his Kingdom of 409. To him it was "a courageous citadel, an outpost of grace, the appointed rallying place for faith, love, beauty, romance and memories. Over it all there presides the spirit of my Jean, now 'gone to singing and to gold' but still a vivid presence, more radiant than ever."

At this time he began writing what he called his "Breviary of Love," a combination of literary anthology and memories, a safety valve for him and a salute to his wife.

Bill Bean bade him not to worry about his lack of energy. "Somehow I have the feeling that entering into communion with you is, in significant ways, like the uplift one gets from experiencing a cathedral like Chartres. Since I know effort is a burden, don't bother replying to this, but every now and then a card to say things are as they are and I hope well enough."

Scarlett did his best, but in some cases the cards or letters were a year apart. He seldom left his house now, except for a short walk with his dogs when the weather was good. Always a man of straight carriage, he had put on weight and had acquired a "reader's stoop." By 1980, although his mind was still clear, he recognized he was suffering from loss of recent memory.

He wrote Bean in August 1980:

The cord of life grows thinner. I am back on a carefully contrived schedule within which I find diversion and exercise. My friends have rallied around and my dear paradigm of a housekeeper is blessedly at the helm. I can only hope old Time has put on his velvet slippers and does not ride roughshod over me. Most of all I deplore the fact that I must curtail writing. It takes too much out of me. I will shortly be limited to writing notes for the milk delivery man, the ultimate in basic English . . . Through this ordeal I never before realized more forcibly that for any one whom the Fates have destined to live in the sparse company of octogenarians, life consists of countless efforts to close ranks and maintain the amenities of the old order, guard the cherished good life, and at all times cling to memories.

To those memories he managed to cling. He was no longer able to play the recorder but he had his records. There was Mrs. Willans to bring him his meals, friends like Smitty Gardner to scour the city for his blessed tobacco — becoming more and more difficult to find — and to take his clothes to be mended, and others who dropped in to chat. There were visits and news from his children and his grandchildren, now grown to eight in number and all of whom he loved dearly.

His Kingdom of 409 and memories kept him alive.

Then, in 1982 his health rapidly deteriorated and on the fourteenth of June Earle Parkhill Scarlett died.

The notice in the obituary column of the Calgary Herald listed many of his accomplishments. This particular inventory ended in an unusual but fitting fashion, quoting the man himself.

Dr. Scarlett once wrote: "In spite of all that has been set down above, the subject of this inventory is still the bemused boy, sitting in the back row of the cosmic theatre, just as he did long ago in the medical theatre of Toronto Varsity, listening, making notes, and wondering."

POSTSCRIPT

There can scarcely be many examples in literature where
the writer of well-loved books was so exactly what the
reader would hope him to be.
— Christopher Morley,
Sherlock Holmes and Dr. Watson

D R. EARLE PARKHILL SCARLETT DIED only a few days short of his eighty-sixth birthday.

He had been honoured by his profession, by his city, by the worlds of academe, literature, history and music. He appreciated those honours but was not overawed by them. He knew, as one of his favourite authors, Thomas Browne, once wrote, "there is no felicity in what the world adores."

Those who read his work – and he had an international coterie of devoted readers – knew him well whether they realized it or not. As Morley wrote concerning Dr. Watson, Dr. Scarlett was as he wrote. Some enjoyed his columns because of their style, that of a person who both knew and loved the English language; but the width of his readership suggested much more than his mere proficiency as a wordsmith. His columns reflected a love for history and literature, founded on his love for medicine. That was the engine which drove his other pursuits.

Time had not, in his final years, been good to him. As he wrote in a private journal after his wife's death: "Time is given too much credit. I hardly agree with the compliments paid to it. Often as not it is a spoiler. Certainly it is not a great healer as far as I am concerned. It is a very indifferent and perfunctory one, and more often than not it does not heal at all."

It was death, not time, which healed his wounds.

Earle Scarlett would not recognize Lake Louise today. As for his Kingdom of 409, the house had to be sold. The buyers tore it down and replaced it with a large modern house which is out of character with the neighbourhood.

But his real kingdom was elsewhere – with his children and beloved grandchildren, with his wife, on the Island of Cos, in libraries, on hospital wards. He was a man of many parts who never forgot that he was, first and foremost, a physician dedicated to that "long" art.

His fears about the future of medicine often seemed an overreaction to the state of affairs as he found them, but that, too, was a mark of his dedication to his principles.

He had written that "what we called 'the humanities' in medicine will not disappear as long as there is one physician left who has not bowed the knee to Baal or who continues to hold the ramparts against the Philistines."

To his death Earle Scarlett never abandoned his post on those ramparts.

NOTES

CHAPTER 1

1. Unpublished interview with Dr. E.P. Scarlett by Charles G. Roland, M.D. (University of Calgary Medical Library Historical Archives), p. 4. Hereafter cited as Roland.
2. Ibid., p. 5.
3. Ibid.
4. *Compendium '78* (Yearbook of E.P. Scarlett High School, 1978), p. 91.
5. Roland, p. 25.
6. Ibid., p. 22.
7. *Compendium '77*.
8. Ibid., p. 7.
9. E.P. Scarlett, "The Medical Jackdaw," *Group Practice*, April 1969, p. 57.
10. Ibid, p. 58.
11. *Calgary Herald*, 15 May 1954.

CHAPTER 2

1. Roland, p. 13.
2. Scarlett's war diaries, Scarlett Collection, Glenbow Archives.
3. Jean Odell's diaries also form part of the Scarlett Collection, Glenbow Archives.
4. Private letter, Earle Scarlett to Jean Odell. This and following letters cited are from the Scarlett Collection, Glenbow Archives.
5. Roland, p. 80.
6. Ibid., p. 81.
7. Ibid., p. 82.
8. Travel diary, Scarlett Collection, Glenbow Archives.

CHAPTER 3

1. Roland, p. 14.
2. Ibid.
3. Ibid, p. 75.
4. Ibid, p. 18.
5. Private letter, Earle Scarlett to Jean Odell, 13 August 1920, Glenbow Archives.
6. Private letter, Earle Scarlett to W.B. Bean, M.D., 21 April 1972, Bean Collection.
7. Scarlett Papers, Glenbow.
8. Dr. A.W. Rasporich, "Medical Man for All Seasons," *Citymakers* (Calgary: Historical Society of Alberta, 1987), p. 176.
9. Roland, p. 30.
10. Ibid.

CHAPTER 4

1. *Alberta Medical Bulletin*, February 1955, p. 114.
2. Roland, p. 26.
3. D.L. McNeil, *Medicine of My Time: A Biographical Interview*, ed. Andras K. Kirchner (Calgary: privately printed, 1983), p. 40. Hereafter cited as McNeil.
4. Ibid.
5. E.P. Scarlett, M.D., "The Medical Jackdaw," *Group Practice*, January 1963, p. 44.
6. G.D. Stanley, M.D., "Daniel Stewart Macnab," *Calgary Associate Clinic Historical Bulletin* 16, no. 1 (May 1951): 19. Hereafter cited as *Bulletin*.
7. G.D. Stanley, *Fun in the Foothills* (privately printed, 1949), p. 35.
8. Harold N. Segall, *Pioneers of Cardiology in Canada* (Willowdale, Ont.: Hounslow Press, 1988), p. 117.
9. McNeil, p. 62.
10. Ibid., p. 32
11. Ibid., p. 34.
12. Ibid.
13. Roland, p. 66.

CHAPTER 5

1. McNeil, p. 52.
2. Roland, p. 33.
3. McNeil, p. 52.
4. Ibid.
5. Roland, p. 34.
6. Ibid, p. 35.
7. Ibid, p. 39.
8. Richard M. Hewitt, M.D., *The Physician-Writer's Book* (Philadelphia: W.B. Saunders Company, 1957).

9. *Canadian Medical Association Journal* 122: 822, 1980.
10. *Calgary Associate Clinic Historical Bulletin* 13, no. 1 (May 1948): 7.
11. *Bulletin* 19, no. 1 (May 1954): 32, 34.
12. Roland, 42.
13. *Bulletin* 1, no. 1 (May 1936): 1.
14. "Odyssey of Medicine," *Bulletin* 1, no. 1 (May 1936): 3.
15. Ibid.
16. W.B. Bean, "Earle Scarlett – An Appreciation," *Canadian Medical Association Journal*, November 1972, 1113.
17. Earle P. Scarlett, "The Ram's Horn" (unpublished anthology in the author's collection), p. V.
18. *Bulletin* 10, no. 4 (February 1946): 167.
19. *Bulletin* 8, no. 1 (May 1943): 1
20. *Bulletin* 9, no. 3 (November 1944): 53.
21. Ibid.
22. *Bulletin* 21, no. 4 (February 1957): 93.
23. Roland, 67.
24. *Bulletin* 22, no. 4 (February 1958): 223.
25. Ibid., 267.
26. Ibid., 268.
27. Roland, 68.

CHAPTER 6
1 Ronald Burt De Waal, *The World Bibliography of Sherlock Holmes and Dr. Watson* (New York: Bramwell House, 1972), p. 303.
2 Christopher Morley, *Sherlock Holmes and Dr. Watson* (New York: Harcourt, Brace and Company, 1944), p. 19.

CHAPTER 8
1 McNeil, p. 69.
2 Ibid., p. 76.

CHAPTER 9
1. Earle Scarlett, "The Medical Jackdaw," *Group Practice*, September 1960, p. 699.
2. Ibid.
3. Ibid., October 1960, p. 798.
4. Ibid., December 1960, p. 967.
5. Ibid., February 1961, p. 130.
6. Ibid., April 1961, p. 277.
7. Ibid., November 1961, p. 860.

8. Ibid., December 1961, p. 953.
9. Ibid., August 1963, p. 499.
10. Ibid., September 1963, p. 580.
11. Ibid., November 1963, p. 698.
12. Ibid., p. 702.
13. Ibid., January 1964, p. 41.
14. Ibid., July 1964, p. 456.
15. Ibid., July 1969, p. 67.

CHAPTER 10

1. *Archives of Internal Medicine*, January 1962, p. 170. Hereafter cited as *Archives*.
2. Ibid., p. 171.
3. Letter, W.W. Bean to E.P. Scarlett, 3 July 1953, Bean Collection.
4. Letter, Scarlett to Bean, 30 January 1957.
5. Letter, Bean to Scarlett, 16 August 1961.
6. Letter, Scarlett to Bean, 29 August 1961.
7. Ibid.
8. Ibid., 25 September 1961.
9. Ibid., 24 January 1962.
10. "Doctor Out of Zebulun," *Archives*, January 1962, p. 171.
11. Ibid., March 1962, p. 160.
12. Ibid., p. 162.
13. Ibid.
14. Letter, Scarlett to Bean, 6 June 1962.
15. "Doctor Out of Zebulun," *Archives*, December 1962, p. 136.
16. Ibid., March 1963, p. 166.
17. Ibid., June 1963, p. 189.
18. Ibid., February 1964, p. 186.
19. Letter, Scarlett to Bean, 11 July 1963.
20. Ibid., 21 September 1963.
21. Ibid.
22. "Doctor Out of Zebulun," *Archives*, 25 March 1965, p. 351.
23. Ibid., p. 352.
24. Letter, Bean to Scarlett, 2 September 1965.
25. Letter, Scarlett to Bean, 24 September 1965.
26. Ibid.
27. Letter, Bean to Scarlett, 23 August 1966.
28. Ibid., 11 November 1966.
29. Letter, Scarlett to Bean, 23 December 1966.
30. Ibid., April 1967.
31. Ibid., 24 April 1968.
32. Ibid., 10 February 1969.

CHAPTER 11

1. Private letter, Scarlett to Bean, 12 January 1968.
2. Ibid., 21 April 1970.
3. Ibid., 15 January 1971.
4. Ibid.
5. Ibid., 11 August 1971.
6. Ibid., 11 August 1972.
7. Earle P. Scarlett, M.B., *In Sickness and in Health,* ed. Charles G. Roland, M.D. (Toronto: McClelland and Stewart, 1972), p. v.
8. Ibid., p. vii.
9. Ibid., p. ix.
10. Ibid., p. xvi.
11. Ibid., p. xvii.

BIBLIOGRAPHY

EARLE PARKHILL SCARLETT

Degrees and Fellowships

B.A. (University of Manitoba), 1916
M.B. (University of Toronto), 1924
Fellow, Royal College of Physicians of Canada, 1932
Fellow, American College of Physicians, 1946
Honorary Doctorate, LL.D. (University of Toronto), 1953
Honorary Doctorate, LL.D. (University of Alberta), 1958
Honorary Doctorate, D.U.C. (University of Calgary), 1969
Corresponding Fellow, Royal Horticultural Society of London, England, 1956–59

Bibliography

The College of Medicine, State University of Iowa. A Sketch of Its History and Its Relations to Medicine in Iowa. (Printed brochure on the occasion of the opening of the new general hospital and medical laboratories building.) University of Iowa, 1928.

The significance of high-grade anemia in chronic nephritis. *American Journal of Medical Science* **178**:215–222, 1929.

Concerning the plague: Some notes on the history of the disease, particularly the great epidemics. *Medical Journal and Record* **130**:647–651, 1929, and **130**:707–713, 1929.

Some aspects of chronic nephritis. *Journal of the Iowa State Medical Society* **20**:167–170, 1930.

Jean Paul Marat: The physician as revolutionist. *Annals of Medical History* (New Series) **2**:71–79, 1930.

Case of acute yellow atrophy (with W.A. Lincoln). *Canadian Medical Association Journal* 26:459, 1932.

The significance of infection in cardiovascular disease. *Canadian Medical Association Journal* 26:562–566, 1932.

Traumatic chylothorax due to intrathoracic rupture of the thoracic duct (with D.S. Macnab). *Canadian Medical Association Journal* 27:29–36, 1932.

The Practical application of our knowledge of the biliary system (with D.S. Macnab). *Canadian Medical Association Journal* 29:281–287, 1933.

The value of administration (with D.S. Macnab). *Canadian Medical Association Journal* 31:489–496, 1934.

Realms of Gold: A Study of John Keats. An address to the University Women's Club of Calgary, 22 January 1935. Privately circulated.

Satira medica: A casual anthology of the satire which has been directed against physicians of all ages. *Canadian Medical Association Journal* 32:196–201, 1935, and 32:314–317, 1935.

Poisoning from phenobarbital (Luminal) (with D.S. Macnab). *Canadian Medical Association Journal* 33:635–641, 1935.

The Odyssey of Medicine. *Historical Bulletin* (Calgary Associate Clinic) 1: 1936. (Note: Dr. Scarlett contributed essays or columns – sometimes both – to each succeeding edition of the *Historical Bulletin*. Columns ran under the title: "A Medical Miscellaney: From the Commonplace Book of a Medical Reader." Because of the sheer volume, titles of essays are not reprinted here.)

An unusual sequence of events in a gastrojejunal ulcer (with D.S. Macnab). *Canadian Medical Association Journal* 37:366–367, 1937.

Functional disturbances of the colon. *Canadian Medical Association Journal* 36:484–489, 1937.

Medicine and poetry: Being an account of those sons of Aesculapius who have on occasion paid homage to the elder god, Apollo. *Canadian Medical Association Journal* 35:676–682, 1936, and 36:73–79, 1937.

A tudor worthy: Master Andrew Boorde of physicke doc tour. *Canadian Medical Association Journal* 38:588–596, 1938.

Medical interlude. Some observations on British medicine with further reflections on the consolations of medicine. *University of Toronto Medical Journal* 16:167–172, 1939.

Masked diseases. *Canadian Medical Association Journal* 41:12–16, 1939.

Three papers on internal medicine read at the 1939 annual meeting of the British Medical Association: Angina pectoris and coronary thrombosis: Clinic study of 100 cases; Common fallacies in diagnosis of cardiovascular disease; Peptic ulcer, new variations on old themes. Published in the *Vancouver Medical Bulletin* supplement for Fall, 1939.

Angina pectoris: A case study. *Canadian Medical Association Journal* 42:34–37, 1940.

"Till the Barrage Lifts." *The Canadian Nurse* 36:476–479, 1940.

Shakespeare's son-in-law: Dr. John Hall. *Canadian Medical Association Journal* 43:482–488, 1940.

Paracelsus. In commemoration of the 400th anniversary of his death. *Canadian Medical Association Journal* 44:510–513, 1941, and 44:618–621, 1941.

William Henry Welch. *Canadian Medical Association Journal* 46:74–75, 1942.

Allergy to liver extract (with D.S. Macnab). *Canadian Medical Association Journal* 46:578–580, 1942.

Silicosis. *Canadian Medical Association Journal* 47:468–472, 1942.

The infernal door: Medical and literary notes on Dr. Jekyll and Mr. Hyde. *Canadian Medical Association Journal* 48:243–249, 1943.

Psychosis in hypoparathyroidism (with W.J. Houghting). *Canadian Medical Association Journal* 50:351–352, 1944.

Five against oblivion. *Canadian Medical Association Journal* 53:391–399, 1945.

Bronchiostasis. *Canadian Medical Association Bulletin* 54:275–283, 1946.

Delta: A problem in authorship. *Canadian Medical Association Journal* 55:299–304, 1946.

Values old and new. *The Canadian Nurse* 44:15–19, 1948.

One hundred years of science in Canada: Medicine, in W. Stewart Wallace (ed): *The Royal Canadian Institute Centennial Volume*. Toronto: Royal Canadian Institute, 1949.

Chronic brucellosis: Diagnosis and treatment. *Canadian Medical Association* 58:230–235, 1948.

The Ram's Horn. (The Mary Agnes Snively Memorial Oration.) Canadian Nurses' Association, August 1948. *The Canadian Nurse* 44:711–719, 1948.

Richard Haydock: Being an account of the Jacobean physician who is also known to History as "The Sleeping Clergyman." *Canadian Medical Association Journal* 60:177–182, 1949.

A Reaffirmation. (Convocation address to the graduating class of the Faculty of Medicine, University of Toronto, 16 June 1950.) *University of Toronto Medical Journal* 28: no. 1, 1950.

Some observations on genius. *Canadian Medical Association Journal* 63:180–185, 1950.

A message from the chancellor. *Evergreen and Gold Yearbook*. University of Alberta, 1953.

Things that abide. *Canadian Medical Association Journal* 68:289–293, 1953.

By Apollo. *Canadian Medical Association Journal* 69:324–328, 1953.

The Invisible Sun. (An address on the occasion of the annual commencement, United College, Winnipeg, 5 November 1953.) Privately circulated.

The Ages of Man. (The Archer Memorial Lecture, Lamont, Alberta, 15 October 1954.) *The Canadian Nurse* 51:607–614, 1955.

The Immortal Memory: Robert Burns. (An address to the Burns Club, 25 January 1955.) Privately printed.

Remarks on the occasion of the unveiling of the portrait of Dr. G.D. Stanley in the Calgary Associate Clinic, 14 November 1954. *Alberta Medical Bulletin*, 32–33, 1955.

Why Study Literature? Broadcast, trans-Canada network, CBC, 10 March 1955.

Variations on a noble theme. *Alberta Teachers' Association Magazine* 35:6–9, 42–47, 1955.

182

A Medical Elixir. (Address to the regional meeting of the American College of Physicians, Regina, 4 February, 1955.) *Manitoba Medical Review*, 499–501, October 1955.

Tangibles and intangibles in medical education. *Canadian Medical Association Journal* 73:85–89, 1955.

Man and the Seven Watchmen: A Commentary on Values in the Modern World. (Address before the 13th Triennial Convention of the Canadian Federation of University Women, Edmonton, 17 August 1955.)

The Many Splendored Thing. (Address to the Canadian Public Health Association, 1955.) Privately circulated.

There was a man . . . I will tell it softly. A tribute to Dr. W.W. Francis, Osler Librarian. *W.W. Francis: Tribute from His friends*. Montreal, Osler Society of McGill University, 1955–56. Edition limited to 500 copies.

Eastern Gate and Western Cavalcade: A commentary on the part played by men of the Montreal General Hospital and McGill Medical School in the early medical history of Western Canada. The third Francis Shepherd Memorial Lecture, Montreal General Hospital, 5 October 1955. Circulated by McGill University. Included in *Canadian Services Medical Journal* 12:851–870, 1956.

The Interpreter's House. (Address at the closing exercises of the University of Alberta, Calgary Branch, 6 April 1956.) Privately circulated.

The Chancellor Reports. (Address to the alumni of the Vancouver Branch of the University of Alberta Alumni Association, 5 May 1956.) Privately circulated.

Tables before You: A Commentary on Food through the Ages. Being the Sixth Annual Memorial Ryley–Jeffs Lecture of the Canadian Dietetic Association, Edmonton, 26 June 1956.

A Walled Stead. (Address on the occasion of the dinner to commemorate the 50th anniversary of the founding of the Calgary Medical Society.) *Alberta Medical Bulletin* 21:3–9, 1955.

Judith Hearne, by Brian Moore. Broadcast, University Station CKUA, December 1956.

Medicine and the Modern Temper. (Being the University Lecture, Queen's University, Kingston, Ontario, 18 February 1957.) Privately circulated.

More Than With a Trumpet. (Address to the Academy of Medicine, Toronto, on the occasion of the fiftieth anniversary of its founding.) *The Bulletin of the Academy of Medicine, Toronto* 30:137–144, 1957.

Salute to a high school. *Yearbook of the Bow Valley High School*. June 1957.

The Dance of Death. (John Stewart Memorial Lecture, Dalhousie University, October 1957.) *Dalhousie Review*, Winter, 1958. Also in *Nova Scotia Medical Bulletin*, January 1958.

The Wisdom to Comprehend. (Remarks at the Initiation Banquet of Alpha Omega Alpha at the University of Alberta, Edmonton, 27 November 1958.) *The Pharos* 22: January 1959.

A Quarry for Wisdom: Some observations on the philosophy of history. Torremolinos, Andalusia, Spain, 1960. Privately circulated.

"The Medical Jackdaw." A monthly column on various subjects printed in *Group Practice* from 1960 to 1971.

A tribute to Gertrude Hall. *The Canadian Nurse* 57: 38–39, 1961.

Fifty years of medicine: 1911–1961. *Canadian Medical Association Journal* 84: 6–10, 1961.

"A Doctor's Notebook." Columns which ran in the *Alberta Medical Bulletin* from vol. 26, no. 2, 1961, to vol. 29, 4, 1964.

Our Medical Heritage. (Address to the students of the Faculty of Medicine, University of Saskatchewan, 20 September 1961.) Privately circulated.

Foreword to the *Nurses' Alumnae Journal*, School of Nursing, Winnipeg General Hospital, 75th Anniversary Number, September 1962.

The chancellor's chair. A note on the origin of the name: Calgary. *The New Trail* 22:7–9, 1962.

What is health? *New Physician* 11:28–29, 1962.

The Interpreter's House. (Address on the occasion of the annual banquet, alpha chapter of Alberta, Alpha Omega Alpha Honour Medical Society, University of Alberta, 1962.) *Bulletin of the Medical Undergraduate Society* (University of Alberta) 6: November–December 1962.

Men for all seasons: Some observations on medicine & society. (Alpha Omega Alpha Lecture, University of Alberta, 24 October 1962. *The Pharos* 26:20–29, 1963.

Round the Fifth Finger. (Address to the 13th Annual Session of the American Association of Medical Clinics, Portland, Oregon, 5 October 1962.) *Group Practice* 12: February 1963.

Report of the Nursing Education Committee (Scarlett Committee). Government of Alberta, 1963.

The old original: Notes on Dr. Joseph Bell, the original Sherlock Holmes. *The Baker Street Journal* (Morristown, N.J.) 13 (new series): 205–209, December 1963.

The good physician: A treasury of medicine. (Review.) *Canadian Medical Association Journal* 90:90, 1964.

Medicine and the modern temper. *Health Insurance Plan of Greater New York News* 11: May 1964.

This is the age of the medical renaissance. *Drug Merchandising* 45: April 1964.

"Doctor Out of Zebulun." A column appearing in the *Archives of Internal Medicine* from 1962 to 1969.

The physician's back garden. *Journal of the American Medical Association* 192:112–113, 1965.

In Sickness and in Health – Reflections on the Medical Profession. Dr. Charles C. Roland (ed.) Toronto: McClelland and Stewart, 1972.

Annual contributions to the Dr. E.P. Scarlett High School yearbook. 1972–1981.

INDEX

www.ingramcontent.com/pod-product-compliance
Lightning Source LLC
Chambersburg PA
CBHW031927080426
42733CB00001B/1/J